About Island Press

Island Press is the only nonprofit organization in the United States whose principal purpose is the publication of books on environmental issues and natural resource management. We provide solutions-oriented information to professionals, public officials, business and community leaders, and concerned citizens who are shaping responses to environmental problems.

In 2003, Island Press celebrates its nineteenth anniversary as the leading provider of timely and practical books that take a multidisciplinary approach to critical environmental concerns. Our growing list of titles reflects our commitment to bringing the best of an expanding body of literature to the environmental community throughout North America and the world.

Support for Island Press is provided by The Nathan Cummings Foundation, Geraldine R. Dodge Foundation, Doris Duke Charitable Foundation, Educational Foundation of America, The Charles Engelhard Foundation, The Ford Foundation, The George Gund Foundation, The Vira I. Heinz Endowment, The William and Flora Hewlett Foundation, Henry Luce Foundation, The John D. and Catherine T. MacArthur Foundation, The Andrew W. Mellon Foundation, The Moriah Fund, The Curtis and Edith Munson Foundation, National Fish and Wildlife Foundation, The New-Land Foundation, Oak Foundation, The Overbrook Foundation, The David and Lucile Packard Foundation, The Pew Charitable Trusts, The Rockefeller Foundation, The Winslow Foundation, and other generous donors.

The opinions expressed in this book are those of the author(s) and do not necessarily reflect the views of these foundations.

CLIMATE AFFAIRS

CLIMATE AFFAIRS

A Primer

MICHAEL H. GLANTZ
National Center for Atmospheric Research

ISLAND PRESS
Washington • Covelo • London

Copyright © 2003 Island Press

All rights reserved under International and Pan-American Copyright Conventions. No part of this book may be reproduced in any form or by any means without permission in writing from the publisher: Island Press, 1718 Connecticut Avenue, N.W., Suite 300, Washington, DC 20009.

ISLAND PRESS is a trademark of The Center for Resource Economics.

Library of Congress Cataloging-in-Publication Data

Glantz, Michael H.
  Climate affairs : a primer / Michael H. Glantz.
    p. cm.
Includes bibliographical references and index.
  ISBN 1-55963-918-0 (acid-free paper) — ISBN 1-55963-919-9
(pbk. : acid-free paper)
  1. Climatic changes. 2. Climatic changes—Environmental aspects. 3.
Environmental policy. 4. Human ecology. I. Title.
  QC981.8.C5G62 2003
  551.6—dc21
                             2002156247

British Cataloguing-in-Publication Data available

Printed on recycled, acid-free paper

Manufactured in the United States of America

09 08 07 06 05 04 03   10 9 8 7 6 5 4 3 2 1

Dedicated to

DR. WALTER ORR ROBERTS
First Director
National Center for Atmospheric Research
Boulder, Colorado

Walt Roberts was the reason I came to NCAR to study the climate system from a multidisciplinary perspective. He was instrumental in launching my career in climate-society-environment studies. World renowned as a scientist, Walt spent much of his career fostering interactions between scientists and industry, between physical and social scientists, and among scientists in a wide range of disciplines. *His legacy lives on.*

and

NANCY MIKESELL RILEY
Administrative Assistant to Walt Roberts

Nancy was a true believer in the goals that Walt pursued—from involvement in the activities of the Swedish-based International Federation of Institutes for Advanced Study to the Aspen Institute's Food and Climate Forum. *Her spirit lives on.*

CONTENTS

PREFACE

The audience for this book includes anyone with an interest, even a passing interest, in understanding how climate can affect human activities, the environment, and the workings of the planet as well as how human activities can affect climate. You do not have to be a climate expert, a meteorologist, or a science whiz to understand climate affairs and to learn why it pays to understand the global climate system on which we all depend. Climate affairs issues encompass climate science, impacts, politics, policy, law, economics, and ethics.

The idea to look at climate issues through the notion of "climate affairs" has been prompted by the spirit of the times. That spirit reflects an obvious growing concern about a broad range of climate issues affecting society, as illustrated repeatedly in news stories, broadcasts, and headlines. The catalyst to put the notion into practice, however, was a specific event, the twenty-fifth anniversary conference for the Marine Affairs Program at the University of Washington in 1998. For the most part, the program's graduates are now professionals working in a wide range of ocean-related activities. Their multifaceted expertise stemmed in large measure from having been educated in marine affairs, which encompasses many academic disciplines. It was an impressive sight to watch the interactions of people with quite different experiences and expertise—engineers, lawyers, economists, ethicists, political scientists, and educators, among others.

Before the late 1960s, formal multidisciplinary marine affairs programs did not exist, even though government representatives worldwide had been involved for many years in negotiating comprehensive aspects of what became the "Law of the Sea" (UNCLOS, 1994). During those negotiations, representatives of the various countries with different, often competing, national interests needed expertise drawn from a variety of academic disciplines and not just from the obvious ones—oceanography or international law.

The case of what to do about manganese nodules on the deep ocean seabed provides one example of the need for multidisciplinary activities. To which government or company, if any, do the naturally occurring nodules belong? How might deep-sea mining rights be allocated? Can the engineering technology be developed to mine them in a cost-effective way? Is it worth the effort financially to retrieve them from the deep ocean's floor? To answer such hypothetical questions, expertise was needed by political negotiators from the fields of economics, political science, law, geography, history, engineering, marine ecology, fisheries, and so forth.

As another multidisciplinary example from the marine environment, human activities have greatly degraded the ecological health of coastal zones. A 7,000-square-mile "dead zone" has appeared in the Gulf of Mexico, centered off the coast of Louisiana (Simpson, 2001). It results from chemical pollutants in runoff making their way into the Gulf from farms and feedlots in the Mississippi River basin (Simpson, 2001). This basin drains about 40 percent of the continental United States. This part of the Gulf has become seasonally hypoxic, which means that living marine resources cannot survive there. Another dead zone has developed in the Caspian Sea off the coast of Baku, Azerbaijan, as a result of raw sewage that is carried by various rivers into the Caspian and from pollution from onshore and offshore oil exploration and processing.

To prepare future decision makers to address such multidisciplinary problems, the first of several academic programs devoted specifically to marine affairs was developed at the University of Rhode Island in the late 1960s. Today, there are several programs that are similar in spirit if not in title. An international association for such programs, Marine Affairs and Policy Association, was founded in 1992 to promote education and research in marine affairs and policy.

It appears that government representatives from around the world are now in the midst of a similar multidecadal process to develop what can be viewed as an international "Law of the Atmosphere." Canadian Prime Minister Brian Mulroney mentioned this idea in Toronto in 1988 at what some observers consider to have been the first international political conference to focus on climate change: The World Conference on the Changing Atmosphere: Implications for Global Security (WMO, 1988). At the time, governmental concern was developing about the possibility of a human-induced global warming of the earth's atmosphere if societies failed to reduce their emissions of certain heat-trapping (greenhouse) gases. More recently, Najam (2000) noted that "international policies for managing the global atmosphere have evolved in an ad hoc and piecemeal manner. It may now be time to adopt a more thought-out and holistic approach. Although some might consider it outrageous heresy, a case can be made for moving toward a comprehensive 'Law of the Atmosphere.'"

A warming of the earth's atmosphere would surely have as yet unidentified effects on local to global climate behavior (i.e., on climate averages and anomalies, weather extremes, and the natural flow of the seasons). Societies and cultures have adjusted over time to their existing regional climates, but altered climate conditions over relatively short periods of time would mean that new socioeconomic adjustments would be needed, some of which could be drastic. Many societies would likely be unable to cope, at least in the short term, with such a major shift in climate conditions, especially during the transition period from one climate type to another.

In the meantime, the number of both domestic and international regulations and laws relating to various aspects of changes in the chemical composition of the atmosphere has been growing. These focus on air pollution, acid rain, transboundary atmospheric pollution, stratospheric ozone depletion, nuclear testing in the atmosphere, and a ban on using the atmosphere for purposes of war. Thus, the educational as well as practical value of developing the notion of climate affairs has become increasingly apparent to educators, decision makers in various socioeconomic sectors of society, and political leaders.

This book provides an overview of climate-society-environment interactions. Place-based case studies are cited throughout to provide readers with examples of the interactions among climate, society, and environment. Collectively, the cases underscore the point that societies everywhere are constantly engaged in coping with variable climate conditions. Each week, scores of new climate-related situations occur, each one of which merits its own distinct discussion and assessment and which yields its own set of "lessons learned." Each year, new climate and climate-related records are being set. The cases cited here illustrate various kinds of impacts, interactions, and problems generated by a variable atmosphere. Decision makers use climate information in conjunction with other nonclimate inputs. In any given situation, climate information may or may not become the major determining factor in a decision because other socioeconomic, political, and cultural factors can and often do heavily influence decision makers at a given point in time. In other words, they are constantly operating in a multistressed policymaking environment.

People interested in climate issues do not need to understand the inner workings of computer models to understand why they are used or to evaluate the output they produce and the limitations of their use for real-world (i.e., operational) decisionmaking purposes. The public can at least evaluate the reliability of the research output of a climate model when it is headlined by the broadcast or print media. Through knowledge of climate affairs, the public will also find itself in a better position to evaluate the implications of conflicting public policy statements made by scientists to the media about the variations, extremes, and impacts of climate. Although many such eye-catching media statements today center on global warming, the public also needs to know about numerous shorter-term climate-related issues.

Examples of climate-society-environment interactions have been placed in various sections of the book based on my perceptions about their primary lessons or illustrations. For example, discussions of Hurricane Mitch and its impacts in 1998 could center on scientific, political, economic, cultural, or ethical aspects. This case focuses on the ethical and equity aspects because, in my opinion, that is the main point I chose to highlight. However, this particular case study could have been used as an example not just of climate ethics but also of climate politics, climate economics, or climate impacts. Also, some examples are used more than once to draw attention to different but equally significant aspects of why it is so important for policy makers, the media, and the public to improve their baseline understanding of the climate system and society's role within it.

Many references are provided to enable the readers to go deeper into the issues raised in this primer. The goal was to prepare a short book to introduce readers to the many ways that climate can, does, and may in the future influence human behavior, and vice versa. Climate-society-environment interplay is too important to be left to the climatologist or, for that matter, to any single set of discipline-focused researchers.

In the ensuing decades of the twenty-first century, the ability of societies around the globe to cope with climate variability, weather extremes, and the likelihood of global warming and its unknown beneficial as well as adverse effects will increasingly be tested and will likely dominate the decision making concerns of national leaders. In this regard, it seems that the twenty-first century has a good chance of becoming "the climate century," a century in which climate-related concerns will occupy significant attention of the next generations of policy makers.

Michael H. Glantz
March 2003
National Center for Atmospheric Research[*]

---

[*]The National Center for Atmospheric Research is sponsored by the National Science Foundation.

ACKNOWLEDGMENTS

Climate affairs, the concept and the book, has greatly benefited from the support of D. Jan Stewart. She has shepherded this manuscript from the earliest days when it was first written down on a napkin in a local diner to the final page proofs. I would venture to say that, after 24 years as administrative and editorial assistant, she knows as much about climate in its multi-disciplinary context as most researchers. Considerable support was also provided by Anne Oman. She identified and redesigned many of the supporting graphics and figures. Thanks also go to Heather Gasper, a very capable research assistant.

Several reviewers of various parts of the manuscript also deserve special thanks: Neville Nicholls (Australia), Clive Spash (UK), Qian Ye (China and ESIG visitor), and Robert Chervin (USA). Todd Baldwin, editor at Island Press, has been a constant support for this activity, and Chace Caven, production editor, who exhibited considerable patience in dealing with the author.

I want to express my deep appreciation to Clifford Jacobs (NSF/ATM), James Buizer (NOAA/OGP), Michael Crow (former Director of Columbia University's Earth Institute), and Kenneth Broad, University of Miami, and Zafar Adeel, United Nations University, for the confidence they placed in the notion of climate affairs well before the manuscript was written.

A final acknowledgment must go to my wife Karen Lynch for her unending support and saintly patience throughout my career, and to Roy and Nancy Lynch of Abingdon, Virginia, who provided me with a summer place to write in Emerald Isle, North Carolina.

INTRODUCTION

ARE WE ENTERING "THE CLIMATE CENTURY"?

Historian Arnold Toynbee once wrote that "in the fifteenth century the Portuguese invented a new kind of sailing ship that could keep to the sea continuously for months on end. This invention suddenly gave the West European peoples the command of the oceans" (1963, p. 2). Portugal was said to have dominated much of the 1500s as a result of its explorations, settlements, and outposts in Brazil and along the lengthy coastlines of Africa, the Indian subcontinent, Southeast Asia, and China.

If you go to the Internet to search for "the Dutch century," "the British century," and so forth, you will pull up a variety of views about which century was dominated by which country. Today, some people argue that the twentieth century has been America's century. According to one source,

> It was in 1941 that Henry Luce [founder of *Time* magazine] first used the term "the American Century" to describe the 20th century, but that term seems even more appropriate now after the end of World War II and the Cold War. During the 20th century, the United States has moved into international politics in a big way, having been the deciding factor in both World Wars and in the Cold War (Glossop, 1999).

Not all writers agree on which century could be claimed by which country because it is a subjective exercise with each observer relying on his/her own set of indicators. Some rely on indicators of conflict, others on technological advances, and still others on social, cultural, economic, religious, or political factors. For example, equally credible arguments could be made that countries such as the Soviet Union or the People's Republic of China dominated international affairs in the twentieth century because of their worldwide ideological influence. Others might say that either Japan or Germany was dominant because their aggressions changed the course of world history. Interestingly, "French historian and political scientist Raymond Aron walked through the ruins of Berlin in 1945, and is reported to have said, 'this could have been Germany's century'" (Smyser, 1999).

Already, at the onset of a new century, suggestions are appearing that this century will likely be dominated by China:

> The 21st century will be the Chinese century, according to more than a few commentators. "Rubbish!" I hear you cry. At the turn of the last century who would have predicted the rise of the USA and the global decline of British influence? Yet historians tell us that the nineteenth century was the British century and the twentieth the American (RCGP, 2001).

Time editors devoted a special issue (26 August 2002) of the magazine to the notion of a Green Century. This was a basic pitch to their readers to pursue activities and make decisions that would be friendlier to the global and local environments. However, from the vantage point of the onset of the twenty-first century, one could argue that this new century is not likely to be dominated by any single country, personality, religion, ideology (including environmentalism), or conflict. Following several decades of rapidly growing governmental and public concern about costly, record-setting meteorological extremes worldwide, along with mounting evidence of a human-induced warming of the earth's atmosphere, we have entered a century in which climate factors are likely to dominate the attention of policy makers. It is foreseeable that this century will be the "climate century."

The process of globalization of communications and transportation, not to mention the ever-growing level of global economic interdependence, assures us that the world's leaders will have to take notice of adverse climate and climate-related impacts wherever and whenever they occur. At the least, leaders will be made aware of the devastation, and many countries will be called upon to supply increasing amounts of financial and food assistance to those countries facing climate-related emergencies. Economic interdependence heightens the likelihood that both adverse and beneficial impacts of a climate anomaly will be felt around the world.

The climate system, too, is a global integrator of sorts, in that local sources of greenhouse gas emissions affect the temperature of the atmosphere globally. A projected warming of that atmosphere could yield surprising changes in the location, frequency, and magnitude of meteorological extremes. Thus, we may find that the twenty-first century will be dominated by memorable variations and changes in climate and increasingly extreme natural hazards. Although the new globalized system is likely to prove more resilient in some respects, it also makes us more aware of just how vulnerable all societies remain to the vagaries of climate.

More and more, media headlines worldwide are about climate- and weather-related disasters:

- devastating hurricanes such as Hugo, Andrew, Fran, Floyd, Georges, and Mitch,
- unexpected European floods (1993 and 1995) and flooding along the Mississippi River (1993),
- destructive rains in Kenya (1997),
- a severe El Niño (1997–1998),
- a devastating ice storm in northeastern Canada and the United States (1998),
- deadly rainfall-induced mudslides in Venezuela (December 1999),
- severe droughts in central and southwest Asia (1999–2001),
- deadly floods in Mozambique (2000 and 2001),
- a cold and snowy North American winter (2001), and

- record-setting droughts and wildfires in various parts of the United States (summer 2002)

and so on. Almost overnight, it appears that societies have become concerned about unprecedented, unpredictable, or unexpected climate-related impacts and fear the future twists and turns that even "normal" climate behavior might take.

Obviously, climate will not be the only issue to reach the media's headlines throughout the twenty-first century. Major issues such as the deadly terrorist attacks in 2001 in New York City and Washington, D.C., the war on terrorism, conflicts in the Middle East, and similar regional conflicts elsewhere have crowded climate issues, such as President George W. Bush's rejection of the Kyoto Protocol, off the front pages—at least for now. However, taking a long view, headlines about climate issues, such as global warming, extreme droughts and floods, El Niño events, loss of climate-sensitive biodiversity, potential famines, energy consumption, and infectious disease outbreaks will continue to appear more and more frequently throughout the century. In the meantime, societies are becoming much more aware of their interactions with and dependence on their natural environments. This is growing into an awareness of their dependence on global as well as local climate regimes.

Reviewing the 1900s

The turn of a century, let alone the turn of a millennium, is a good time to reflect on the century gone by and to speculate about the new one. So far, there have been scores of retrospectives written on the wars, technological and health breakthroughs, international politics, and economic, social, and environmental changes in the 1900s. To that list one can add the numerous publications and videos that reviewed the weather and climate conditions of the past 100 years (see appendix on Twentieth Century Climate Extremes.) Some of these reviews provide a chronology of the improvements in weather and climate technologies, our understanding of the global climate system, and the numerous ways that human activities can influence climate on different time and space scales.

Many people believe that industrialized countries are less

vulnerable to weather and climate extremes than others, but that is an untested assumption. Rich industrialized countries as well as relatively poor developing ones have to date been unable to weatherproof or climate-proof their activities to totally avoid climate-related damage to life and property. Nevertheless, numerous technological breakthroughs of the twentieth century occurred with regard to weather forecasting, climate prediction, and reducing societal vulnerability to extreme meteorological events.

Over the course of the century, for example, people modified their weather and climate conditions in various ways at local to global scales. Scientists pursued cloud seeding to increase precipitation, especially during drought episodes. Many attempts were made to bring rainfall to arid and semiarid lands by modifying the land surface to affect atmospheric flow patterns (e.g., plant trees, pave the desert with asphalt strips). Deforestation in the Amazon is known to have an adverse impact on the amount of rainfall in the rainforest because half of the precipitation in the region comes through evaporation from the basin's vegetation. When air conditioning technology was introduced, it was seen by economic development specialists as a way to export temperate climate conditions to tropical areas, making the tropics a much more hospitable place (from a European perspective) for settlement and development by colonial powers.

With the end of the Cold War in the early 1990s, considerable media, government, and research attention shifted to climate and weather issues: recurrent El Niño and La Niña events, global warming, heat waves, floods, tropical storms, droughts, fires, and frosts. Also in the 1990s, the hottest decade in the historical record, new temperature and precipitation records were set each year somewhere on the globe, and we witnessed an increasing intensity of extreme events such as tornadoes, hurricanes, and typhoons. Each year now seems to bring with it weather- and climate-related surprises that had neither been witnessed nor anticipated by scientists or decision makers.

Although throughout the twentieth century governments had already engaged in a wide range of efforts to avoid the adverse effects of climate, many societies became, in many ways, more vulnerable to climate problems than ever. Climate had traditionally been defined in statistical terms that centered on the physical aspects of the atmosphere. By the end of the century, however, a narrow definition was no longer very descriptive of the climate system or useful for society's needs. Societies need more information about climate than just its numerical characteristics. Information about climate has become extremely important as societies try to cope with the issues they care about: food, water, energy, health, safety, commerce, communication, recreation, travel, and so forth. Each generation of decision makers needs to be reeducated about the interactions between climate and human activities and be updated on new climate-related trends, extremes, and fluctuations. Population increases and migration into previously uninhabited areas drive an increased need for climate-related information. In the last century, human settlements encroached on floodplains worldwide, expanded into arid areas and pristine forests, worked their way up unstable hillsides and mountain slopes, and filled in previously uninhabited sections of vulnerable coastal areas. Thus, not only does the climate change on various time scales but so too does the social climate.

Climate Roulette at the Beginning of the Twenty-First Century

In a given region, the perception of the climate and its actual characteristics are two separate matters. People expect the climate to behave in certain, often desirable, ways without surprises. Those perceptions, based on what they have seen or heard, however, often do not match reality, and as a result, the coping strategies devised by society are frequently challenged by the reality of climate and weather extremes. In some regions, even average conditions have their unexpected impacts. For example, in an arid area, average climate conditions are not necessarily a good thing. In fact, average conditions seldom occur. It is important to understand perceptions of climate because actions taken based on them will have real consequences for society, even if those perceptions are incorrect.

Many people view climate variations, and especially extremes, as some sort of adversary. They think something is out to get us, and we must control climate variations by any means possible. Nevertheless, some would argue that the climate is neutral; it exists regardless of what societies do. As societies developed, most likely at first in environments that were generally hospitable, human activities changed in situ and expanded into areas that were likely to have been increasingly less hospitable. Government policies, population increases, a few wet years or decades, and the need to support families even at a subsistence level have prompted people to move into areas considered marginal from the standpoint of rain-fed food production and sustainable development. Growing the types of foods that different societies value, but in inappropriate locations (e.g., in areas with irregular or insufficient rainfall), can lead to lower than anticipated crop yields, food shortages, and even famines.

The pressure to exploit increasingly marginal areas means that climate-related problems are likely to increase. Although the climate system has often been blamed for food production problems, many of the problems relate to human choices and perceptions of acceptable risks. From the perspective of society, however, climate variability and weather extremes are like predators that can generate adverse, unpredictable impacts on environment and society.

Anomalies in the climate system can also be considered predators from the perspective of natural ecosystems. For example, coral bleaching is associated with both El Niño events and global warming (i.e., the increase in the average sea surface temperatures that global warming is predicted to bring). Plant and animal populations are influenced periodically by extreme variations in regional climate characteristics and by climate change over the long term.

For the most part, a large portion of the world's population is held hostage by climate. People's lives are governed by local rainfall and temperature conditions that combine to determine the characteristics of their growing seasons. Many cultures have adjusted their activities to the expectable vagaries of the climate system where they live. They have done so by choosing crop types that can prosper in local climate conditions, by developing ways to bring water to fertile but otherwise dry soils, or by raising livestock where the rainfall is too erratic to sustain crop production from year to year.

Interestingly, many of society's attempts to mitigate the impacts of weather and climate were designed not to resolve climate-society issues but to circumvent them. For example, instead of trying to match crops to existing climate conditions, methods were devised to circumvent climate constraints:

- irrigation in arid and semiarid areas,
- high yield varieties (HYVs) in areas with relatively poor soils,
- genetic manipulation of crops,
- greenhouses to create desired climate conditions for plant life,
- chemical fertilizers to improve crop yields,
- increased water quantity by building dams and reservoirs in arid regions.

However, simple solutions to existing problems often generate their own new sets of environmental problems. HYVs require major inputs of water and fertilizers and are less tolerant of large variations in temperature and precipitation. Irrigation can lead to the salinization and waterlogging of soils. The use of marginal soils often leads to extreme soil degradation and even to desertification.

Each of these problems poses a difficult challenge to society, and their growing frequency and intensity leads to ever higher risks. The risks may ultimately include the very survival of civilization or cultures. From this perspective, the way that various societies gamble with climate looks very much like a game of Russian roulette.

Climate Affairs in the Twenty-First Century

In the twenty-first century, climate science will most likely advance slowly with occasional major breakthroughs. Climate anomalies and their impacts on societies will likely increase. Societies will continue to collect lessons learned from the impacts of successive climate-society interactions. The value and use of climate and climate-related information will increase but at an uneven pace. Population growth rates and movement into increasingly marginal areas are expected to increase the vulnerability of settlements to climate variability. People in the United States, for example, are moving toward the coasts and putting themselves increasingly in harm's way of tropical storms and storm surges. As the uninhabited spaces along the coastline have filled with settlements, hurricanes no longer make landfall without causing major damage.

Countries will continue to face variable climate conditions on seasonal and interannual time scales. El Niño events will appear every few years or so. There will be decades-long shifts in climate regimes, as had been the case throughout the 1900s, for example, the decline in flow of the Colorado River from about 18 million acre-feet (maf) in the early 1920s to 13.5 maf for the rest of the century. In the first several decades of the twenty-first century, the ability of societies around the globe to cope with such climate variability, weather extremes, and the adverse and beneficial effects of global warming will increasingly be put to the test.

This book is intended to outline a new field called climate affairs, which seeks to address the societal need for a better understanding of the many ways that variability, change, and extremes in climate affect ecosystems and the affairs of people and nations. Armed with such knowledge, decision makers in all socioeconomic sectors of society will be better informed, as will the general population, about climate-society-environment interactions. Climate affairs must include consideration of climate science, policy, law, politics, economics, ethics and equity, and climate's impacts on ecosystems and societies.

Chapter 1 presents definitions and perceptions of climate and aspects of climate behavior. Seasonality, which is an underplayed but extremely important characteristic of climate, is highlighted. Ecosystems, societies, and individuals around the globe live, for the most part, according to the natural flow of the seasons. A bridge between concerns about weather and climate is identified.

Chapter 2 notes that climate-related issues are already acknowledged as important to societies and governments. That importance continues to grow, given heightened societal sensitivity worldwide to climate variability and the likely impacts of global warming on human activities and the environment. Here, the case is made that the twenty-first century will not be dominated by any particular country or ideology but by the behavior, especially anomalous behavior, of the global climate system.

Although people may be aware of the climate influences in their own part of the world, chapter 3 explains how no part of the earth really escapes the climate system's extremes. In addition, all parts of the globe will be affected by global warming and all coastal areas will be affected by any rise in sea level.

The notion of "climate affairs" and its various aspects are discussed in chapter 4. Chapter 5 provides several examples of societal uses, misuses, and potential uses of climate-related information, including forecasts.

Chapter 6 addresses the problems associated with different climate impact research methods. The notion of foreseeability is discussed in terms of its potential value for forecasting climate-related impacts to allow society to act before devastating impacts occur. The concern is also addressed that present-day students of the atmosphere are only dimly aware of the large volume of previous research that exists on climate, weather, and climate impacts. The works of previous generations of researchers should be mined for new ideas about how societies are affected by as well as influence climate on all spatial scales.

The concluding chapter highlights some important but overlooked aspects of climate-society-environment interactions. Interest in the results of those interactions is the *raison d'être* for fostering the notion of climate affairs as a multidisciplinary approach to an improved societal understanding of the workings of the global climate system. The appendix reviews the twentieth century's major climate events and their consequences.

Sustainable Development

This book describes a conceptual framework and then its component parts. This approach was used to build a strong, credible case with a foundation for understanding why climate matters to all societies, rich and poor, industrial and agrarian, sedentary and nomadic, North and South, democratic and authoritarian. The various societies and parts of the earth are affected in different ways by global climate behavior, as atmospheric changes are influenced by a wide range of geophysical characteristics. I also wanted to make clear that humans are now a notable forcing factor for climate at the global as well as local level. Human activities not only change the climate but also make societies and ecological systems more vulnerable to its variations. This is why climate affairs ultimately belong under the rubric of sustainable development, as a field that crosses disciplinary boundaries to provide insights into a more fundamental pursuit: to improve the quality of life and environment for current and future generations. The Brundtland Commission (WCED, 1987) originally defined sustainable development as "the capacity to meet the needs of the present without compromising the ability of future generations to meet their own needs."

The concept of sustainable development now has scores of definitions and websites (e.g., www.ulb.ac.be/ceese/meta/sustvl.html) ranging from a focus solely on economic factors to a broader focus on socioeconomic and quality-of-life issues. As a result, "sustainable development" can be taken to mean almost anything one wishes. Any given definition of the concept has its supporters (e.g., the International Institute for Sustainable Development, www.iisd.org) and its challengers (e.g., Redclift, 1987). When it comes to setting policies or attacking any particular problem, government leaders can always find a "credible" excuse for deviating from one path of sustainable economic development in the name of another. Thus, even those who profess to be devoted to sustainable development raise questions about the notion.

For example, is California a good example of sustainable development or is it sustaining itself by exploiting the resources of other regions (e.g., water and energy), impinging on their drive toward sustainability? One California writer (Barris, 1999) made the following observation about agricultural activities in his state: "Unsustainable agriculture is farming with another region's water and doing harm to that other region." Southern California is an arid and therefore water-short region. It has been drawing an excessive amount of water from the Colorado River system. Water is also channeled to the south from the northern part of the state. Energy is drawn from outside the state, as well as from elsewhere within its borders. During a recent drought in California, the governors of Alaska and

California proposed that the federal government support an $80 billion undersea pipeline to transport water from a water-rich environment to a water-poor one. The question remains: Is California an example of sustainable development? If so, at whose expense?

Clearly, climate variations on the time scales of concern to a given generation influence a country's efforts to achieve sustainable economic development, however it is defined, over the long term. However, climate affairs interface with sustainable development in not so obvious, but still important, ways. A high-impact weather or climate anomaly may last only a few hours, days or even a season, but its impacts can linger for years, diverting scarce resources toward reconstruction efforts and away from development activities. Any policy devised under the umbrella of sustainable development must take into account its short- and long-term impacts on the local, regional, and global climate system. Otherwise, a boomerang effect can result from seemingly good but nevertheless misguided efforts.

Historical, contemporary, and futuristic examples have been provided throughout this book to integrate the various components of climate affairs. These examples show how the various interacting features of a changing atmosphere and human activities combine to produce environmental and societal impacts with which decision makers must contend, in addition to all the other stresses confronting them. Even governments whose only interest is in sustaining themselves politically must at least feign an interest in sustainable development.

ONE

WHAT IS CLIMATE?

All peoples, regardless of their level of economic or political development or type of political system, are subjected to the vagaries of weather and climate, for good and for ill. They always have been. Today, however, the focus of interest in climate and weather seems to center on their adverse impacts on society and the environment. We see news articles about devastating droughts such as those in sub-Saharan Africa in the 1970s and 1980s, in Australia in the early 1990s, and 2002, and in central, south, and southwest Asia (specifically Uzbekistan, Afghanistan, and Iran) from 1999 to 2001, as well as extreme flooding in Central Europe in the summer of 2002. We are also beginning to see strong interest in industrialized and agricultural countries in the influence over long distances, called teleconnections, of air-sea interactions in the tropical Pacific Ocean (i.e., El Niño and La Niña events). According to Myers and Kent, "among geophysical agents, weather and climate are by far the most lethal to humankind worldwide. Together flood, hurricane and drought account for 75 percent of the world's natural disasters. Only earthquakes exact a comparable total" (Myers and Kent, 1995).

Governments feel obligated to respond, but, as the saying

goes, everyone talks about the weather (and now the climate), but no one does anything about it. Adding to the pressures on governments and individuals of having to cope with the impacts of climate variability is the increasing likelihood of a human-induced warming of the earth's atmosphere. Such a warming is largely a result of increasing amounts of carbon dioxide emissions that result from the burning of fossil fuels (coal, oil, and natural gas). As late as the 1980s, arguments could legitimately be put forth that the then $0.5°C$ ($0.9°F$) atmospheric warming was within the bounds of the twentieth century's natural variability. However, by the beginning of the twenty-first century, those arguments had weakened considerably because the 1990s and beyond have been among the hottest years in more than a century of observations. Most recently, 2002 was identified as the second hottest year on record.

James Hansen, director of the NASA Goddard Institute for Space Studies, testified in a U.S. congressional committee hearing that he believed the costly 1988 drought in the Midwest was a sign of human-induced global warming (Hansen, 1988). Reagan's high-level political advisers were far from convinced that global warming was caused by human activities. Those advisers immediately challenged newspaper reports of Hansen's testimony as politically unacceptable because they were counter to the policy of the Reagan administration (Begley, 1988). Some scientists (e.g., Trenberth et al., 1988) also challenged Hansen's linkage of the 1988 drought to global warming. By the late 1990s, however, many researchers, politicians, and media were suggesting publicly that "weird" weather extremes and climate happenings after 1988 were likely to have been, as Hansen had suggested earlier, manifestations of global warming.

Interest in the effects of climate is not limited to global warming. The onset of the twenty-first century provides a new opportunity for people to learn just how much their lives are affected by variability and other changes in the global climate system. It is not very difficult to argue that the lives of everyone on earth are increasingly entwined with average climate conditions, especially how those average conditions may be changing, and how they are occasionally punctuated by anomalous climate and weather extremes.

Each year, somewhere on the globe, a drought, flood, fire, killing frost, or climate-related disease outbreak is catalyzing devastation and misery at the same time that other locations are producing bumper food crops and have favorable water supplies. However, extremes and hazards have a regional bias. For example, Californians do not worry about the hurricanes that annually occur in the Atlantic and plague the East and Gulf coasts of the United States. For their part, people in Florida are not concerned about the severe coastal storms and heavy rains that tend to hit southern California during El Niño events. Those who live in the southern United States do not concern themselves with a severe winter in the northern half of the country. Inhabitants along the east and west coasts, although sympathetic, do not worry very much about tornado outbreaks that occur in the Midwest every spring.

Most people subscribe to the following: if you don't like the weather where you live, wait a few days and it will likely change. If you don't like the climate where you live, move! However, the reality of a relocation is that you will have traded a set of climate hazards in one region for a different set of hazards elsewhere. Moving to Florida from Oklahoma means moving to a region threatened by hurricanes from one plagued by tornadoes. This could be referred to as the localization of concern about regional natural hazards, or the "mesoscale problem."

Defining "Climate"

According to Upgren and Stock (2000), "weather is a visible and tangible manifestation of the physical conditions of the air at a given location and time, and to the change in these conditions." Upgren and Stock noted that the atmospheric conditions of weather can be measured using only four parameters: temperature, pressure, dew point, and precipitation (p. 7).

Climate is a statistical creation used to represent a variety of atmospheric conditions. To many, climate is viewed as average weather. However, climate is not just a number averaged over a certain period of time from months to decades or over a given area. Statistical averages hide from decision makers important information about regional climate. Agricultural and other government planners can be misled if they rely solely on average annual rainfall numbers. Depending on location, different degrees of variability (i.e., departures from average) in both time and space exist for precipitation, temperature, cloud cover, relative humidity, wind speed and direction, and so on. For example, rainfall is highly variable in arid areas; a few high-rainfall episodes are averaged with a larger number of well-below-average rainfall events. During a 40-year period in the West African Sahel, the average annual rainfall amount was within 10 percent of the long-term average only twice (Glantz and Katz, 1985).

Averages are only one way to describe the climate of a particular region. One could, for example, refer to normal climate conditions (the climate conditions that seem to prevail most of the time) or to the modal climate condition (the type of climate that prevails most frequently). Each of these measures may differ from the way people perceive the climate conditions in a given area, which may not accurately reflect reality.

Strictly defined, climate can also vary on all space scales from the very local to the global level. It can vary within a country at the same time, with one location exhibiting below-average conditions while another exhibits above-average conditions in precipitation and/or temperature. Because climate is a statistical representation of weather conditions in a given place, people can measure weather characteristics in their own backyards. The weather conditions will most likely be similar to those witnessed in neighboring backyards. The farther away one goes, the more likely it is that the weather and, therefore, climate conditions will be different because of different amounts or types of vegetation, the type of cloud cover, or topographic features such as differences in elevation. Relatively local conditions can also change in different parts of the region without changing the regional climate's average conditions.

Most people are not really well versed in the distinction that researchers make between the concepts of climate and weather. As a result, these words are frequently used interchangeably. This lack of distinction is constantly reinforced by the media's use, or misuse, of these terms. Too often, scientists themselves use these terms interchangeably, even though they know the definitional distinctions between weather and climate. Scientifically accepted definitions of climate and weather are

> *Climate.* [A]verage meteorological conditions over a specified time, usually at least a month, resulting from interactions among the atmosphere, oceans, and land surface. Climate variations occur over a wide range of spatial and temporal scales.

> *Weather.* [A] condition of the atmosphere at a particular place and time measured in terms of wind, temperature, humidity, atmospheric pressure, cloudiness, and precipitation. In most places, weather can change from hour to hour, day to day, and season to season. (Burke et al., 2001, pp. 113–114)

When physical science researchers refer to climate information, most likely they are referring to atmospheric data (called time series) and, more generally, to statistical information about weather conditions averaged over various lengths of time for specified areas. However, when a social scientist refers to climate information, he/she is referring to those aspects, but also to information about climate and weather impacts on societies and ecosystems, climate- and weather-sensitive decisionmaking processes, the level of vulnerability of a society in the face of climate anomalies, and even societal responses to forecasts. Although meteorologists and climatologists try to keep weather and climate considerations separate, one could argue that weather extremes are climate-related and, therefore, fair game for research by climate impacts researchers.

> [W]eather is what is happening to the atmosphere at any given time, while climate is what we would normally expect to experience at any given time of the year on the basis of statistics built up over many years. But when it comes to discussing the impact of extreme events, this distinction is less easy to maintain. (Burroughs, 1997, p. 3)

One additional concept is important to understanding talk about climate: the phrase "climate-related" refers to second-order or indirect impacts of climate on societies and ecosystems.

This includes, for example, the impacts of drought on crop yields and production, on water resources, and on changes in the location of disease-bearing vectors. Third-order impacts of drought emanate from the second-order ones, as a drought's impacts ripple through society. Such "knock-on" effects include increased food prices as a result of reduced crop yields, water rationing or reduced irrigation because of water shortages, reduced hydropower production and resultant brownouts and blackouts, and unanticipated outbreaks of infectious diseases among humans and animals. The knock-on effects of a drought readily translate into a degradation of environmental quality. The sum total of all climate and climate-related impacts has greatly increased interest in society about the many ways that global climate influences just about all aspects of human activities and ecological processes.

The Climate System

The climate system is presented as a recurrent theme throughout the book. As a result, considerable attention is focused on the atmosphere and on processes, both natural and human, that can influence its composition and behavior. This might give a reader a false impression that other components of the climate system are not as important in the workings of the system. That would be unfortunate for, as with the human body, all of the components are important to the maintenance and viability of the body as a system, and each component of the climate system has a key role to play, albeit some components are more crucial than others. While focusing on the atmosphere and climate in the Primer, please keep in mind that these are only some of the parts of the Earth's climate system. In a way, it is analogous to focusing discussion on the heart in the human body, as opposed to the whole body in order to identify its functions and influences on other parts of the system.

The global ocean is a key component of the climate system for a variety of reasons. The oceans make up 71% of the earth's surface. It stores heat and carbon. It has a longer-term memory than does the atmosphere. It retains heat for longer periods of time than the air above it. Its surface and subsurface processes influence the climate at regional to global scales. Its level rises and falls on all time scales. It is the source of many climate-related hazards (hurricanes, cyclones, typhoons, El Niño and La Niña events, tsunamis, storm surges, salt-water intrusions, changes in the thermohaline circulation) that affect ocean currents, internal waves, warm ocean currents such as the Gulf Stream and the Kuroshio Current, upwelling, and so forth. Clearly, oceanic processes and atmospheric processes are entwined, often in ways that are difficult (as yet) to identify, such as which one leads or lags the other in time and space. The oceans are in a dance, so to speak, with the atmosphere.

Given the average of weather features for a specific region and time, an anomaly would be any deviation from the average. Although they may not be seen as such by a large portion of the public, many weather and climate anomalies are normal parts of the climate regime. Sometimes, however, the deviations from average conditions are so large that they capture the attention and concern of the public and policy makers. How one decides what is an anomalous but normal event, or anomalous and unusual, has a lot to do with one's perception of this type of weather event experienced in the past, as well as its frequency, magnitude, and location. Before we talk about the aberrations of climate, however, we need to address what constitutes a "normal" climate.

Various aspects of the climate system have long intrigued the public and scientific researchers. The climate system serves as the common foundation for all the other aspects of climate affairs. There are many scientific and popular sources of information about the climate system, both highly technical and very general (e.g., Houghton et al., 2001; Suplee, 2000).

A traditional representation of the global climate system highlights the system's major interacting physical components: the cryosphere (ice), the biosphere (vegetation), the hydrosphere (oceans and other water bodies), and the lithosphere (the earth's crust). Human activities are a recent additional component of the earth's climate system.

The earth receives energy from the sun. Because of our planet's spherical shape and its orbit in the solar system, different parts of the earth's surface receive varying amounts of solar energy;

the tropics receive the most and the polar regions the least. Because of this differential heating at the earth's surface, winds and ocean currents are generated that serve to redistribute energy between the tropics and the poles on different time and space scales (figure 1.1).

The reflectivity of the earth's surface (called albedo) determines in large measure the amount of incoming solar radiation that is reflected back into the atmosphere and space or is absorbed so as to heat up the earth's surface. Thus, the extent of light-colored surfaces on the globe such as ice and sand and dark-colored surfaces such as forests is crucial to the planet's balance of radiation.

Evaporation from the oceans, vegetation, and soils provides the moisture to the atmosphere that leads to cloud formation. The rising motion of the atmosphere in the tropics as a result of heating creates clouds, whereas the descending motion of air in the subtropics as it cools stops cloud formation. This is a major reason for the belt of deserts around the globe. Evaporation is greatest in the tropics. The heating up of the oceans creates considerable water vapor. As evaporative processes increase, so does the amount of water vapor, a greenhouse gas, which then makes a warm atmosphere even warmer.

Incoming solar radiation heats up the earth's surface, and longwave radiation is reemitted from the surface of the ocean and land. Some of the radiation is intercepted by clouds and reflected back toward the earth's surface, further warming the land and oceans. This is referred to as the greenhouse effect and is a naturally occurring process. The earth's lower atmosphere is further warmed as a result of gases and aerosols that have heat-trapping characteristics, the so-called greenhouse gases (water vapor, carbon dioxide, methane, nitrous oxide, and chlorofluoro-carbons [CFCs]). All but the CFCs are naturally occurring gases. Without these gases, the temperature at the surface of the earth would be about 0°F (−18°C), as opposed to its current average of about 59°F (15°C). The greenhouse effect has made possible the existence of life on our planet.

The climate system is also influenced by occasional volcanic eruptions that emit large amounts of particulates into the upper atmosphere where they intercept and reflect back to space

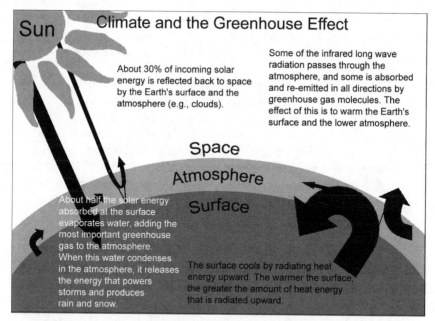

Sun Climate and the Greenhouse Effect

About 30% of incoming solar energy is reflected back to space by the Earth's surface and the atmosphere (e.g., clouds).

Some of the infrared long wave radiation passes through the atmosphere, and some is absorbed and re-emitted in all directions by greenhouse gas molecules. The effect of this is to warm the Earth's surface and the lower atmosphere.

Space

Atmosphere

Surface

About half the solar energy absorbed at the surface evaporates water, adding the most important greenhouse gas to the atmosphere. When this water condenses in the atmosphere, it releases the energy that powers storms and produces rain and snow.

The surface cools by radiating heat energy upward. The warmer the surface, the greater the amount of heat energy that is radiated upward.

Figure 1.1. A schematic of the earth's Greenhouse Effect.

additional amounts of incoming solar radiation before it can reach the surface of the earth. This leads to a temporary cooling of the atmosphere.

Carbon dioxide emissions have greatly increased as a result of the burning of fossil fuels and tropical deforestation. Tropical rainforests are carbon "sinks" (i.e., they take carbon dioxide from the atmosphere). Cut down a forest and a carbon sink has been removed, although it becomes a sink again if it is allowed to regenerate. Since the onset of the Industrial Revolution in the mid-1700s and the associated steady increase in the use of fossil fuels, the atmospheric content of the major greenhouse gas, carbon dioxide, has increased from an estimated 270 parts per million by volume (ppmv) to the 2001 level of over 371 ppmv.

Today, scientists refer to greenhouse gases in terms of their global warming potential, that is, how much each one contributes to the increase in global temperatures. In the past the climate system was viewed as consisting of only physical and biological aspects. We now know the extent to which human activities can produce changes in the terrestrial and atmospheric environments that tend to enhance the naturally occurring greenhouse effect.

Human activities also emit greenhouse gases other than carbon dioxide to the atmosphere. Nitrous oxide is used in fertilizers for agricultural production. Methane seeps into the air from landfills and from livestock. CFCs were manufactured for use as refrigerants, foam-blowing agents, electronic cleansing solutions, and as propellants in spray cans.

A human-induced warming of the atmosphere by a few degrees Celsius will change regional climate regimes by altering temperature patterns (e.g., warmer nights, warmer winters), precipitation patterns (more or less precipitation, depending on location), and the characteristics of the seasons. Climate scientists have made a wide range of educated guesses about how and where the temperature, precipitation, and seasons might change; a band of uncertainty accompanies each of these "guesstimates."

Systematic observations of rainfall and temperature are of relatively short duration with respect to the history of the earth, as are direct measurements of atmospheric gases. To compensate for this, there is a reliance on paleostudies to provide indirect (called proxy) information and on computer simulations to identify past climates and, based on certain assumptions, to gain a glimpse of possible future ones. For example, ice cores taken from the Antarctic and from various glaciers have helped to provide information about climate changes over the past several hundred thousand years. Geological assessments that focus on changes in the earth's surface layers also provide insights to paleoclimates and their anomalies, such as floods and droughts.

Although a brief description of the climate system cannot do justice to the intricacies of the structure and function of that system, it can provide the reader with a place to start for a deeper understanding of the system.

Climate Behavior

"Normal" climate can be viewed in at least two ways. For a given space, it can be defined in terms of the actual climate characteristics, such as amount and type of precipitation, temperature values, changes in humidity, cloud cover percentages over time, and so forth. These are measurable factors, and because they can be put into statistical terms, "normal" can be objectively defined. This is very useful information for many decisionmaking purposes: where to locate a ski or beach resort, what kind of clothing to manufacture and when to have it in retail stores, how big to build reservoirs for water supplies, and so on. Risks of outlying extreme events can be calculated, and they too can be objectively and quantitatively defined using the best available information. Anyone can go to an almanac, encyclopedia, or the Internet and see what the normal climate is for a specific city or country.

A second way to view "normal" has little to do with objectivity or perhaps even reality. "Normal" is most often defined as a result of perceptions. People tend, for example, to weigh recent climate conditions more heavily than past conditions. As a result, what they consider to be normal climate is really based on a set of conditions witnessed over a relatively short period of time—in other words, a statistically meaningless small sample. Hence, their perceptions are likely to be an imperfect reflection of reality. Different people doing different things for work or for recreation, and living in the same area for different lengths of time will likely hold different views about what constitutes the normal climate and what might be viewed as anomalies.

In an arid or semiarid area, for example, many observers previously viewed drought as a departure from the normal climate. The truth is that droughts in any region are part of the normal climate. Frequencies, intensities, and magnitudes may change over time. Nevertheless, droughts as anomalies in precipitation are part of the normal climate and are to be expected. However, with an absence of drought for a lengthy period of time, it disappears from the perception of what constitutes normal climate.

Because people in different parts of the globe are not necessarily subjected to the same set of climate problems, it is important for the reader to be aware of the wide range of ways in which climate anomalies can affect human activities, directly as victims and indirectly in the variations in cost for energy, food, and fiber products. Such awareness is also important for humanitarian purposes.

Climate behavior encompasses the following: *climate variability* from season to season and from year to year; *climate fluctuations* from decade to decade; *climate change*, long-term major changes in the average global temperature and precipitation (also referred to as "deep climate change"); *seasonality*, the natural flow of the seasons; and *extreme meteorological events* (e.g., droughts, floods, fires, frosts, and severe storms).

Climate Variability

Atmospheric processes vary on all time scales, and as a result, climate varies from seasons to years to millennia. For many of a society's decisionmaking purposes, variability refers to seasonal and interannual time periods. Over centuries and millennia, scores of civilizations in different environmental settings around the globe have had to learn, often the hard way, about how to cope with the natural flow of the seasons and the variability in climate from year to year. These are the time scales at which societies, individuals, and governments alike prepare their food, water, energy, health, and even defense needs. Planting schedules, for example, are obviously based on seasonal considerations of temperature and precipitation. Responding to their perceptions of local seasonality, farmers have learned through experience when it is best to plant and harvest their crops, when to weed, when to watch for pests, how to recognize water stress in crops, and so on. Farmers have learned from their experience and the experiences of their predecessors how not to "fight to bankruptcy" the flow of the seasons or interannual variability but how to capitalize on them. For example, farmers in developing countries often keep some seed grain in reserve in case of

---

BOX 1.1.

Anomaly

anomaly (1664) 1 : deviation from the common rule : irregularity 2 : the angular distance of a planet from a perihelion as seen from the sun 3 : something different, abnormal, peculiar, or not easily classified

—Webster's Ninth New Collegiate Dictionary

drought. This provides them with enough grain for food and planting in the following year. To know that variability occurs is one thing. To forecast it from season to season and year to year is quite another matter.

Climate variability manifests itself in various ways. For example, the succession of years of either deficient or surplus rainfall is often overlooked. In the West African Sahel, as in other parts of the world, farmers can often withstand a year of drought. Some can even withstand two consecutive years. Very few, if any, however, could survive a third year of drought, as occurred in the early 1970s. Even in a rich country, farmers have great difficulty surviving a few years of drought. The point is that in many societies the sequencing of wet and dry years makes a difference: wet followed by wet evokes one type of response, dry followed by dry another type. Likewise, wet, dry, wet is likely to evoke a different response than a dry, wet, dry sequence. Remember that a year or two of drought can cause problems for an economy that last several years beyond the end of the drought.

Every so often, the global climate system is perturbed by another manifestation of variability—El Niño. El Niño as a *basinwide* phenomenon involving air-sea interactions was discovered by an atmospheric scientist in 1969 (Bjerknes, 1969). El Niño events recur on average every 4.5 years, but the range can be between two and ten years, as suggested by figure 1.2.

El Niño is an occasional warming of sea surface and below surface temperatures in the central and eastern equatorial Pacific Ocean. It disrupts seasonal climate patterns to varying degrees around the globe for twelve to eighteen months, sometimes longer. This phenomenon has occurred for thousands of years (Diaz and Markgraf, 1992).

Although its influence for the most part is worldwide, El Niño–related disruptions of normal climate conditions appear to be more predictable for some areas than for others. For example, its influence is the strongest in the tropical and extratropical Pacific Ocean (i.e., Pacific Rim countries and islands) and around the tropical belt that girdles the globe. Individuals and policy makers are now showing interest in and want to learn more about how forecasts of El Niño (and La Niña, anomalously cold

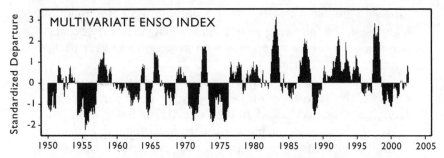

Figure 1.2. Time series of the El Niño-Southern Oscillation using the multivariate ENSO index, which is based on the six main observed variables over the tropical Pacific. These six variables are sea-level pressure, zonal and meridional components of the surface wind, sea surface temperature, surface air temperature, and total cloudiness fraction of the sky. El Niño (0 to +3), La Niña (0 to –3). (Source: CDC, 2002)

sea surface temperatures in the central Pacific) might be used to prepare for the likely impacts of natural hazards spawned by these phenomena. Global and local response strategies must take into account the uncertainty in El Niño forecasts before reliable coping mechanisms can be developed. Researchers are very interested in this phenomenon because they strongly believe that it can be forecast reliably. Forecasts notwithstanding, the onset of El Niño can be identified in real time thanks to an elaborate ocean monitoring network called the TOGA-TAO Array, a system of buoys that monitors changes in sea temperatures across much of the equatorial Pacific.

Societal awareness of and experiences with El Niño events began to develop only in the mid-1970s, and interest in La Niña only in the mid-1990s. Understanding El Niño is similar to the way we tend to understand the seasons: we know they will return but are not sure whether they will be severe or mild, wet or dry, early or late. Unlike with the seasons, however, no regular return period exists. Nevertheless, knowing that El Niño is a recurrent phenomenon provides societies with some possibilities to think about in preparing for its impacts.

Climate Fluctuations

Bryson and Murray (1977) asked, "Why do we refer to 'climate fluctuations' rather than 'years with adverse climate'?" They then explained:

> Many of our food crops are very well adapted to the climate under which they are grown—for example, corn. This means that the best yield is obtained under climatic conditions like those of the recent past, often called "normal" climate. Any departure from "normal" brings a lower yield; any fluctuation is adverse for well-adapted crops (Bryson and Murray, 1977, p. xii).

Global and regional climates also vary on decadal time scales. The term "fluctuations" refers to decadal-scale climate variability; the concept received widespread use during the 1970s. It appears to have lost favor in the 1980s and 1990s, perhaps giving way to the phrase "climate change." The popular term for fluctuations among scientists now is decadal-scale variability. CLIVAR, the Climate Variability and Predictability Program, was established by the scientific community in the mid-1990s to bring research attention back to decadal-scale changes (fluctuations) in the climate system. However, it would be useful to bring the term "fluctuations" back into the climate-impacts vocabulary because it helps bridge the conceptual gap between short-term climate changes (seasonal, annual) and the longer term changes (centuries and longer). These distinctions could help the public, the media, and political leaders to better understand the nuances of variations in the global climate regime.

In the fifty-year run of the El Niño-Southern Oscillation (ENSO) cycle shown in figure 1.2, the mid-1970s appears to have been a dividing line. Before then, there had been a greater number of La Niña events than El Niño events. Afterwards, however, the opposite has been the case. This suggests fluctuations in the dominance of one ENSO extreme over another.

The history of Atlantic hurricanes provides another example of climate fluctuations. Hurricanes were more frequent in the 1944–1967 period than between 1968 and 1991. The 1995–2000 hurricane seasons have been labeled as the most active on record to date. Some hurricane researchers suggest that the natural fluctuations in hurricane frequency mean that there will be a return period on the order of decades of a higher than average number of hurricanes in the region, as had been the case in 1944–1967. As a result of this reasoning, they contend that more Atlantic hurricanes in a given year in the near future may not be the result of global warming but would be just a manifestation of natural climate fluctuations (Gray et al., 1997).

Yet a third example of fluctuations relates to water resources and the changing amount of annual flow in the Colorado River. Until the early 1920s, flow in this river system had been estimated at about 16 million acre-feet per year. In retrospect, what was considered normal flow between 1900 and 1920 was actually relatively high. From the early 1920s to the 1980s, however, the annual streamflow dropped considerably (figure 1.3). It is not yet clear whether Colorado River flow will increase again to its previous high levels as a manifestation of regional natural fluctuations in the climate system.

Climate fluctuations also leave their mark on ecological processes. For example, a severe drought in West Africa in the late 1960s and the 1970s aroused concerns in the United Nations and area governments about the negative short- and long-term effects of desertification—processes that create desertlike landscapes. This occurred along the southern edge of the Sahara Desert, a region known as the Sahel, during that time.

Satellite images and ground photos showed desert sands on the move into populated regions. Images from space showed large volumes of Saharan dust being transported across the Atlantic basin and into the Western Hemisphere. Blame for this widespread degradation of Sahelian vegetative cover and the appearance of "deserts on the march" shifted from natural to human causes and finally to a combination of the two. In the late 1980s, however, a review also based on satellite imagery of the Sahelian zone suggested that desertification processes along the southern edge of the Sahara had not only stopped but in some locations had even been reversed (Tucker et al., 1991; Pearce, 2002). This environmental change had occurred despite increased pressures on the land's vegetative cover resulting

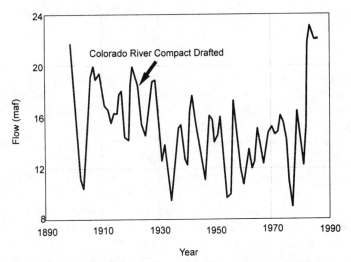

Smoothed Flow vs. Year

Figure 1.3. Reconstructed streamflow at Lee's Ferry, Arizona. The data have been smoothed using a three-point smoothing algorithm called hanning. (After: Brown, 1988)

from human activities (farming, herding, woodcutting). The review by Tucker and colleagues suggested the existence of a natural decades-long ebb and flow of climate-influenced desertification processes in this region. It also suggested that speculation about the long-term impact on soils and vegetation of human activities in this and other fragile arid and semiarid regions should not be based solely on seasonal or year-to-year observations.

In sum, one could argue that climate variability is the aspect of climate that most individuals and societies are concerned about and respond to—a mild winter, a cool summer, an anomalously wet or dry spell, an intense tropical storm. Whether the climate fluctuates over decades or changes in fundamental ways in the long term to a warmer global average state, seasonal to inter-annual variability will still occur with the newly established global average conditions. Most individuals and societies take a much shorter term view of climate and weather conditions because those are the time scales with which they have to cope.

Climate Change

A major problem exists in communicating to the public the meaning of the phrase "climate change." To most people on the street climate change is mixed up with climate variability. "This summer was wetter [or hotter] than last summer," is an example of a common popular reference to climate change. One might hear someone say, "When I was a child, I remember that there were more hurricanes than there are today." When scientists talk of climate change, they are referring to shifts in the global climate system that have not been witnessed for one or more centuries or even millennia, changes the likes of which societies have no memory.

In an attempt to get around the confusion about what constitutes a climate change that is considered by scientists to be really profound change as opposed to variability on time scales of direct interest to societies today, the former can be labeled "deep climate change." This would help to draw a long-needed line of separation between those climate "changes" that we have been living with and have adjusted to over seasons, years, and even decades, to those that societies have not witnessed in centuries. Many scientists now say that future generations will likely need to adjust to this profound type of change as a result of constantly increasing greenhouse gas emissions.

Deep climate change refers to a warming of the earth's atmosphere that is expected to range between 1.5 and 5.8°C (2.7–10.5°F) by 2100, as projected by general circulation modeling output. The models also project an average sea level rise of 0.09 to 0.88 m by the end of this century, depending on the model scenario used (IPCC, 2001). A climate change of 0.6–0.7°C (1.1–1.3°F) occurred in the course of the twentieth century. Although some portion of this increase appears to be human-induced, many changes of various magnitudes have occurred in the past as a result of natural factors. As noted earlier, the impacts on the atmosphere of these anthropogenically emitted gases, along with alterations of the land's surface (deforestation, cultivation, and grazing activities, urbanization, desertification) and industrial processes are considered by many scientists to have caused the global average temperature to have risen (see Santer et al., 1996).

To geologists and climatologists, climate change also refers to long-term climate swings that have taken place in the distant past and will likely occur in the future. Aside from climate changes millions of years ago, changes of interest to us today include more recent periods of cooling and warming: the Younger-Dryas (a cool period about 11,500–12,800 B.P.), the Altithermal (a warm period about 6000–10,000 B.P.), the Medieval Optimum (a warm period about 1000–1200 B.P.) and the Little Ice Age (1500–1850 B.P.). Researchers today warn about abrupt climate change, suggesting that deep changes could take place in a matter of decades (Overpeck et al., 1996; NRC, 2002).

Most people do not understand the possible environmental consequences of a relatively small (a degree Celsius or two) and seemingly insignificant increase in the global atmospheric temperature. Nevertheless, scientists are monitoring for signs of temperature change in many geographic locations from pole to pole. Of special interest are marginal climate areas such as the transition areas between climate zones—the cold margins in the polar regions, the dry margins along desert edges, and the high margins in mountainous areas. High-altitude and high-latitude regions supply some of the most convincing evidence that the global climate has been changing: glaciers around the globe have been retreating for several decades. If, for example, the ice floating in the polar seas were to melt, global atmospheric changes and therefore climate changes would occur because less solar radiation will be reflected back to space.

In the late 1970s, some scientists pointed to one of the most dangerous potential impacts of global warming, the possible disintegration of the West Antarctic Ice Sheet (WAIS). If ocean temperatures were to warm significantly in the southern ocean that encircles Antarctica, it would undermine the stability of the WAIS, much of which rests on land. If a large segment of the ice sheet were to collapse into the ocean, it would cause a very rapid rise in *global* sea level by several meters. This would inundate many island states in the South Pacific and Indian Ocean, as well as many coastal urban centers around the globe. Although WAIS has a low probability of disintegrating, that probability is not zero. When it comes to sea level rise, there would be no climate-change winners. This is unlike changes in the location and amount of precipitation, where some countries would likely benefit while others would lose.

With a gradual global warming, subtle changes will occur in things like crop yields, production, and composition; the frequency and intensity of droughts, floods, and tropical storms; the volume and timing of snowmelt and stream flow; and the geographic ranges of certain dreaded tropical disease-bearing insects.

Seasonality

People everywhere live by the seasons. For some, the seasons are tied to sports, recreation, or industry; for others, they signal work cycles. For many others, the natural rhythm of the flow of the seasons is a matter of survival. This is true in industrialized economies as well as in economies in transition and other developing societies. A close look at the flow of the seasons and societal dependence on it reveals several subtle ways that seasonality changes can either harm or benefit societies. Only recently has the notion of seasonality begun to receive more attention by researchers in fields other than anthropology and economics (Sahn, 1989).

An interesting example of seasonality at work relates to poverty. In many developing countries, poor peasants borrow seed, funds, or farming implements from local entrepreneurs so that they can grow crops for consumption or sale. They must pay back the lender with crops or labor after the harvest. This arrangement works until there is a drought- or flood-related crop failure. If the borrower cannot pay back the lender in cash or crops, he must pay him back in labor. Therefore, in subsequent years, he must first prepare the lender's land before he can tend to his own fields. His production declines and he becomes increasingly indebted to the lender. This is an example of how a disruption in seasonality because of a climate-related anomaly such as a drought or flood can keep poor people poor while helping the relatively rich become richer. The Independent Commission for International Humanitarian Issues (ICIHI, 1985) commented on the importance of seasons in developing countries: "the essential point is that country life is potentially rich with signals of crisis, but these are seasonal" (p. 40).

Many social, economic, and political factors affect the ability of groups to cope with the societal impacts of seasonality, making it difficult to generalize. Table 1.1 represents the effects of seasonality on a generalized agricultural community. It is presented here to suggest to the reader how the usual rhythm of the seasons might affect the rural population in a community faced with a wet season and a dry season. The second column of the chart suggests how drought might affect the seasonal rhythm of agricultural activities.

The impact of drought on society will vary according to its timing. If it occurs at the onset of the growing season, for example, those farmers in a position to do so might have the chance to replant several times using their seed reserves. If drought occurs in midseason, the number of options available to farmers, such as planting varieties that require shorter growing seasons, would be fewer because the growing season would have already been shortened considerably.

Concern about global warming, and climate change in general, should also increase interest in seasonality because most societies around the globe have adjusted their human activities to expected seasonality at the regional level. Farmers know about when they should begin to prepare their fields for planting, when to plant different types of crops, and when to harvest. Ranchers and herders know when to move their livestock into and out of grazing areas in tune with the life cycle of seasonal vegetation. Manufacturers know when to make winter and summer clothes for retail stores. Fishermen know when and where to fish for different species, and so forth. Water resource managers monitor the amount and water content of snowfall in the mountains to forecast potential stream flow months in advance. Abnormally high temperatures alter the timing and the amount of stream flow. This can greatly upset the planting schedules of farmers who depend on snowmelt water to irrigate their fields.

Table 1.1.
Possible Societal Responses to the Cycle of the Seasons during Favorable and Drought-Plagued Periods

(Usual) seasonal impact		*Prolonged drought impact*
POST-HARVEST (EARLY DRY SEASON)	•food available •food prices decline •migrant laborers return to villages •morbidity declines •mortality declines •nutritional status improves	•food availability declines •food prices continue to rise •domestic food self-sufficiency jeopardized
"Post-harvest food availability largely determines the size and distribution of village calorie supplies, not just at the time but until the next harvest" (Schofield, 1974, p. 23).		•families borrow from kin/friends •disposal of assets for money •migrants do not return •additional family members migrate
DRY SEASON (LATE DRY SEASON)	•food becomes less available •food prices increase •nutritional status (esp. women, children) declines •drinking water becomes scarce •dry season irrigation becomes more important	•nutritional intake deteriorates •morbidity stays high or increases
WET SEASON (EARLY WET SEASON)	•"hunger season" begins •wild food use begins •gathering added to agricultural labor •high food prices •poor families borrow agricultural inputs •distress borrowing/distress sales •draft animals in relatively weak condition •diseases more prevalent •morbidity increases	•food prices go even higher •little work available in rural areas to earn cash •distress sales increase: livestock, stored grain, household goods •food unattainable (due to lack of availability, high price) •food-gathering activities intensify •nutritional status declines •seeds eaten (reduces future production)
"During the wet season itself when seasonal food shortages peaked, hardship could be partially alleviated by participation in communal work parties and short-distance migration making use of the variation in the onset of the rains (and hence in the timing of planting, weeding, and harvest)" (Watts, 1983, p. 49).		•eat plants/leaves, not usually eaten •children's illnesses increase •morbidity increases •mortality increases •irrigation becomes limited as
PRE-HARVEST (LATE WET SEASON)	•food prices are at seasonal high •food intake is lowest (esp. women and children) •body weights decline •"hunger season" peaks	stream flow is reduced •call on wider networks; reliance on more distant kin, and on national and international agencies
"Peak-season labour inputs often coincide with seasonal food shortages, as on-farm grain stocks are running low before prices begin to be pulled down by the impending harvest" (Schofield, 1974, p. 23).		

Column 1 adapted from Chambers et al., 1979.

Extreme Meteorological Events

Extreme meteorological events (EMEs) generate both fear and fascination in societies. Without a doubt, the extremes of weather, climate, and climate-related events capture the lion's share of attention of policy makers, the public, and the media—much more so than changes in statistical averages of climate and weather. Climate hazards that always make the newspaper headlines and television newscasts include droughts, floods, forest and bush fires, killing frosts and freezes, tropical storms, tornadoes, heat waves, cold spells, blizzards, ice storms, and strong winds.

In the late 1990s, Americans heard about the worst windstorms in France in 200 years, and Europeans learned of Venezuela's devastating mudslides. Deadly floods in Bangladesh and China and lengthy droughts in central Asia were news spots instantaneously heard around the world.

Extremes are a constant concern to those who bear responsibility for public safety and economic well-being. Aside from society's interest in the awesome power of nature, extremes tend to remind individuals, corporations, and governments just how vulnerable they are to the extremes of regional and global climates.

Just about every spot on the globe is subject to climate-related hazards. For the most part, though, societies have developed formal and informal institutions and mechanisms that they expect will reduce societal vulnerability before, during, and after an extreme event. In many countries, special institutions or mechanisms have also been set up specifically to ensure societal resilience, that is, the ability to bounce back in the aftermath of climate-related and other natural disasters. These institutions and mechanisms can be used not only on the country's own soil, but in other countries as well.

Climate-related extremes of one type or another occur somewhere on the globe each and every season in varying numbers and with varying levels of damage. A complete list of all climate-related extremes in a given time period would fill many pages. Humanitarian and other donor organizations with limited financial and human resources have relatively little time to deal with one climate-related emergency before they are called on to address another emergency somewhere else.

Societies everywhere are aware of the local and regional risks of facing one or more climate or weather extreme. As a result, EMEs, and climate change-related shifts in their frequencies, intensities, and locations have become central research interests of climatologists. Because of the strong interest in extremes within the climate and weather research communities, EMEs should serve as a bridge between communities that otherwise would tend to focus only on the specific time scales of interest to them.

Through extremes, weather affects people every year, directly and visibly at the local and regional levels. When a flurry of tornadoes accompanies a severe weather front, decimating towns and disrupting lives along its path, people throughout the country empathize with the victims. However, tornadoes, as a constant risk or high threat, fall into a relatively small space and time box in the United States and have relatively few victims.

As former Speaker of the U.S. House of Representatives, Tip O'Neill, once suggested, "All politics are local"; so too could one say that "All weather is local." Local weather phenomena generate the most public interest and concern, especially in the areas that are directly affected by them. Concern, however, diminishes sharply with distance from the affected area and with time as well. Despite local adverse impacts on people and property, elsewhere, a sort of cognitive dissociation from those impacts relegates them rather quickly to a level of insignificance in the unaffected regions. The bottom line is that sympathy for the weather-related problems of others rapidly dissipates. Each person is to some extent concerned about his/her regional or local weather anomalies, but shows much less concern about those in other parts of the country.

Most often people take serious notice of weather forecasts when they are harmed: "Why weren't we warned about the timing or extent of floods along the course of the Upper Mississippi River basin?," they might say. "Why was the forecast for the winter in the Northeast so far off base?" Although people expect so much from their national weather and climate services, they are often not surprised when forecasts fail to match what actually occurs. In many cases, it seems that they expect them to fall short of the mark. Rightly or wrongly, weather forecasters are often the targets of jokes. For example, Patrick Young said, "The trouble with weather forecasting is that it's right too often to ignore it and wrong too often for us to rely on it" (The Quote Garden, 2003).

It seems that society accepts a certain number of weather-related problems as part of life as usual. For example, cities constantly prepare for the problems that they believe are most likely to affect them, such as protecting property from hurricane winds and rain, snow removal, deicing aircraft, and repairing roads (e.g., potholes) each spring. Rather than try to reduce those impacts to zero, people accept them, despite programs to drought-proof, flood-proof, and weather-proof a region. However, when the number of potholes goes beyond an unspecified but perceived threshold, society wants action. In a recent winter in the northeast United States, cities faced three times the normal number of potholes, and their budgets proved woefully inadequate for road repair. Yet, even with a forecast of a severe winter, money must be found to cover the anticipated higher costs of repair. Scarce financial resources would have to come from somewhere else in the budget.

Societies are also susceptible to climate- and weather-related surprises such as the devastating ice storm in January 1998 in northeastern North America. These cannot necessarily be well prepared for in advance of their occurrence. Nevertheless, those societies that do not take weather and climate information seriously are unnecessarily putting themselves at increased risk to variations in the climate system.

Opposing views about the impacts of a weather-related natural calamity underscore the fact that there are different perceptions of responsibility for the severity of the impacts. What adverse impacts can rightfully be blamed on the weather versus on decisions that set societies up for weather-related disasters? For example, flooding in the summer of 1993 in the Midwest along the upper Mississippi evoked opposing descriptions about the same event. A farmer whose land was inundated said, "The Mississippi River overtopped the levees and flooded my farmland." However, what actually happened was that the Mississippi River overtopped the *human-made* levees and flooded the *natural* floodplain, which had been converted to farmland a long time ago. To some observers, this was a weather event believed to be a once in a lifetime occurrence. To other observers, it was a climate extreme that could be expected to occur x times in y years.

Despite valiant attempts to define weather and climate, it appears that people already tend to think about weather and climate in different, often conflicting, ways. The media unfortunately reflects the public's confusion about these concepts. For example, looking up weather and climate in the annual index of the *New York Times* news items, one finds all such listings under Weather along with a note under the Climate heading to "see Weather." The index of the *Times* (London), however, lists all weather and climate news items under the Climate heading with a note under Weather to "see Climate." The *Christian Science Monitor's* index lists weather and climate news stories as separate and independent entries.

Bridging Climate and Weather

Superstorm '93

Extreme events are of common interest to weather and climate researchers alike. Superstorms provide a good example of why researchers who focus on climate change, climate variability, or extreme events need to work closely together. For example, in mid-March 1993, storm cells developed in the Gulf of Mexico, over Mexico, and over the midwestern United States. They later merged into the large storm system that resulted in what became known as "Superstorm '93." The geographic scope of its impacts ranged from Cuba, through twenty-six states in the eastern United States, and into northeastern Canada (figure 1.4).

This was an unprecedented storm for a couple of reasons. Its large magnitude had not been seen in recent times. It caused the drowning of more people than hurricanes Andrew and Hugo combined. It spawned a variety of unanticipated hazards of unexpected intensities (i.e., tornadoes, coastal storm surges, heavy snowfall, blizzards, high winds). According to Fischhoff (1994), despite timely forecasts of this severe storm, "hundreds of people still died from exposure, heart attacks, falling tree limbs, road accidents, and the like. No one counted the number of injuries, illnesses and close calls" (p. 387).

The death and destruction that resulted from this particular

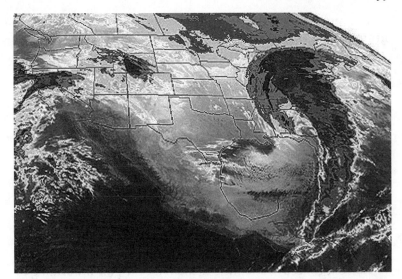

Figure 1.4. Meteosat infrared satellite photo of the 12–14 March, 1993 "Superstorm '93."

superstorm was relatively high (about 250 people were killed), even though a reliable early warning was issued by local and national government agencies about five days in advance of the movement of the storm system. According to a report by the National Oceanic and Atmospheric Administration, "the Superstorm of 12–14 March 1993, was among the greatest nontropical weather events to affect the Nation in modern times." The report also noted that the superstorm had a negative impact on more than 100 million Americans and severely affected economic activities in the eastern third of the United States. Its repercussions were felt across the country, as 25 percent of commercial flights nationwide had to be canceled on Saturday and Sunday (March 13 and 14). Superstorm '93 was estimated to have caused well over $2 billion in damage (NWS, 1994).

This same superstorm also had costly adverse impacts in Cuba and Canada. For example, the Cuban media reported that there were a dozen deaths and millions of U.S. dollars' worth of heavy wind damage to crops and property.

The precursors to the superstorm were well monitored on the ground and from space as they were developing. Official government warnings were issued about the storm and its high potential for damage. The Weather Channel also communicated warnings. Fischhoff (1994), a psychologist and decision theorist, suggested that "obviously, the value of forecasts comes from providing the needed answers in a usable form. Somehow or other, the Weather Channel's 'superstorm' weather forecasts were not answering the questions that viewers were asking, or were not being understood as intended" (p. 387). So warnings from a few credible sources notwithstanding, deaths and damage still resulted. In retrospect, the warnings were not always taken seriously. In some cases, there was little communication between those doing the warning and those receiving them (e.g., local emergency managers).

Current collective scientific wisdom on the global warming of the earth's atmosphere suggests that a warming will likely be accompanied by an increase in the intensity, frequency, and geographic range of regional climate-related hazards. The causes and consequences of Superstorm 93 lend themselves to speculation about how similarly large, destructive storms could develop in places where they have not previously been experienced. Superstorm 93 and other anomalies (such as the supercyclone of October 1999 that made landfall in Orissa, India) can generate discussion of just how well industrialized and developing societies might be able to cope with future storms of similar magnitudes.

El Niño Weather

El Niño (and La Niña) events also provide a good example of where weather meets climate. In some parts of the globe, an El Niño can influence weather conditions in some, if not all, of the seasons during its life cycle. In the United States, for example, people living in the northern half of the country can usually expect a warmer and milder winter than normal. For their part, Southern Californians can expect an increase in the frequency and amount of rainfall. Inhabitants of the southeast can expect to be wetter and cooler than normal in the winter. In northern

Peru, inhabitants can expect heavy rains and flooding, as well as mudslides in the northern Andes. Northeast Brazil can expect hotter weather and less rainfall in the early months of the year following El Niño's beginning, its growing season. The point is that the characteristics of the seasonal weather patterns in El Niño-influenced regions differ from normal weather patterns. It is difficult, however, to blame a specific weather event on El Niño. The occurrence of El Niño alters the probabilities of occurrences of certain types of weather and climate-related anomalies. Hence, it is legitimate to refer to El Niño weather, even though weather researchers have in the past shuddered at the thought of doing so.

Reviewing suggestions about how best to define the concept of climate is an enlightening task. It seems that the scientific concern over distinguishing between climate and weather has generated ongoing discussions since at least the earliest decades of the 1900s. To the public, it may be enough to say that climate is what you expect and weather is what you get. This observation captures the main differences between the two concepts: climate is the average of weather conditions over some period of time, and weather is what is happening now and in the very near term. It also captures the human aspects by referring to perceptions. Perceptions are not necessarily valid reflections of reality, because few people tend to think of climate in terms of its full range of characteristics.

The focus in the next chapter is on the various ways that climate anomalies interact with human activities. It draws attention to the fact that each location around the globe has its own set of climate-society interactions with which governments and individuals must cope. These anomalous aspects of the climate system capture the attention of policy makers, weather-sensitive socioeconomic sectors, and the media. The chapter begins with a brief discussion of how climate might be perceived (as a constraint, resource, or hazard).

TWO

CLIMATE AND SOCIETY

Everyone has heard the question: If a tree falls in the forest and no one is around to hear it when it falls, does it make a sound? The analogy to climate is quite clear: If a climate or weather anomaly does not affect people, would people care about it at all? The fact is that a climate anomaly is noticed when it wreaks havoc in a society. People tend to pay the most attention to those anomalies that impact life and property, and the closer to home it is, the more attention they pay. Thus, people are most directly concerned about climate-related impacts on society and the environment, even though they may show some interest in the physical aspects of the climate anomaly itself.

This chapter provides an overview of the three major but different and conflicting ways that people view climate. It then provides discussion of various climate and weather anomalies that tend to affect societies in adverse ways. The Appendix presents the reader with the view of experts from the US National Oceanic and Atmospheric Administration (NOAA) and from the UN World Meteorological Organization (WMO) about which weather and climate extremes they consider to have been the most notable of the twentieth century.

Three Perspectives about Climate

Climate has been viewed in one of three ways: *climate as a constraint* to social and economic development; *climate as a resource* to be fully exploited to society's advantage; and *climate as a hazard* that can spawn other hazards and disasters.

Climate as Constraint

A few decades ago, a report (Greenwood, 1957) on climate and economic development raised the following concern: "By any rational definition of 'underdeveloped country' most of them are entirely—or partially—in the tropics. Is [a hot tropical] climate the common factor that keeps them underdeveloped?" Many people still consider their regional climate conditions as a constraint on human activities, a boundary condition that they had to accept and adjust to in order to survive or prosper.

One name associated most closely with this deterministic view of climate is that of Yale professor Ellsworth Huntington (1915). He believed, for example, that "a certain type of climate, now found mainly in Britain, France, and neighbouring parts of Europe, and in the Eastern United States is favourable to a high level of civilization. This climate is characterized by a moderate temperature, and by the passage of frequent barometric depressions, which give a sufficient rainfall and changeable stimulating weather" (as quoted in Brooks, 1926, p. 292).

Huntington also suggested that "the climate of many countries seems to be one of the great reasons why idleness, dishonesty, immorality, stupidity, and weakness of will prevail" (Brooks, 1926, p. 204). This reinforced a belief that people were lucky to be born in a favorable climate (in terms of food production, water resources, and energy production) and unlucky to have been born in a hostile climate in the tropics where people have to struggle to eke out a livelihood in large measure because of their "poor" climate setting. Huntington's work became associated with racism because it intimated that people in the tropics were less productive than those in the temperate zone.

For a few decades after Huntington's book was published, it was as if climate could not be discussed in the same breath as economic development. With the exception of relatively few publications, it was not until the mid-1970s, because of the deadly 1968–1973 drought in the West African Sahel, that climate was once again brought back into development discussions. It was, however, still viewed as a boundary constraint for developing countries. Even textbooks about economic development in sub-Saharan Africa had not paid much direct attention to the topic of recurrent drought and its impacts on development processes, even though parts of the continent had just gone through severe drought and famine.

That all changed by the late 1970s, when many articles, books, workshops, and conferences began to appear on the interrelatedness of drought, desertification, and food security issues throughout the African continent. Instead of talking about climate in general, interest shifted to discussions of some of the manifestations of climate, such as droughts and floods. By focusing on its anomalous episodes, climate and development once again became fair game for discussion without the discussants having to fear being labeled racists. Kamarck (1976), for example, bypassed discussing climate as a constraint and hinted at the impacts of climate variability as a factor in economic development in the tropics.

Two interesting references to climate/development issues raise some very basic concerns. I uncovered (not discovered) the first reference by accident in a book (Cox, 1875) on the Crusades. Cox wrote that in A.D. 1095, Pope Urban II at the Council of Clermont, while urging the Church's followers to participate in the Crusades in the Holy Lands, stated that "the blood which ran in the veins of men born in countries scorched with the heat of the sun was scanty in stream and poor in quality as that which coursed through the bodies of men belonging to more temperate regions" (p. 30). This "uncovery" suggested to me that the perceived differences between abilities of inhabitants of the tropics and those of the temperate zone preceded Huntington's belief by about 900 years. This discriminatory belief appears to be more the result of human nature and culture than of colonialism.

A second interesting reference appeared in the early 1980s, when an Indian social scientist commented on the potential effects of tropical climate as a boundary constraint to economic development. Bandyopadhyaya (1983) wrote that "the global climate dichotomy is . . . responsible for the wide and widening economic gap between the North and the South and the neo-imperialistic economic exploitation of the latter by the former. . . . It is evident that the amelioration of the tropical climate is one of the necessary conditions for the total emancipation of the South from the economic and political control of the North" (p. 158). He concluded by arguing that global warming could actually benefit the Third World by reducing the economic disparity that exists between nations of the North and the South.

Climate as Resource

Most of the time in any given location, climate is a resource. Climate is exploited by societies in many ways. We adjust our lives and activities to the expected flow of the seasons. The dictionary defines the word "resource" as "a source of supply or support" and as "a natural source of wealth or revenue." When people think of climate as a resource, for the most part they think of it as good climate conditions for agriculture or livestock grazing. Such a climate provides adequate water supply directly or indirectly in a region where the soils are fertile enough to support the growth of food crops or foraging livestock. The growing season is most often favorable for rain-fed agricultural production or rangeland use.

Some countries or regions have ideal conditions for food-producing activities. A few of them have even been referred to as "breadbaskets," that is, regions where food production is reliable, sustainable, and plentiful. Examples include the United States' Great Plains, the Ukraine, Argentina, the Canadian prairie provinces, and the Awash River basin in Ethiopia. Vietnam used to be referred to as the rice bowl of Asia. There are many more parts of the globe that are, could become, or once were highly productive food-producing regions.

In some cases, these proverbial breadbaskets were destroyed because of regional climate fluctuations or changes, while in other cases inappropriate land-use practices (e.g., overgrazing and trampling of vegetation and soils by livestock, poor irrigation drainage, salinization of soils, wind and water erosion) destroyed the land's productivity, even though the climate did not change.

Today, climate as a resource encompasses much more than the physical components of the climate system. Information about the behavior of the system is a valuable resource to those who have access to it and who know how to use the information. Probabilistic forecasts on daily to decadal time scales provide valuable information for planners in a variety of socioeconomic sectors (e.g., agriculture, water, energy, and health). Advance warnings, however imprecise, do provide at the least a heads-up to planners and decision makers engaged in climate-sensitive activities. For example, wind maps would be invaluable for determining the best places for energy-producing wind farms, as would seasonal precipitation forecasts for agricultural areas.

Another aspect of climate as a resource includes the people who are involved in climate and climate-related research and research application to meet societal needs. Even if a developing country does not possess the homegrown expertise it needs to effectively exploit certain kinds of terrestrial or marine resources, its researchers can call on the assistance of such United Nations organizations as the WMO, the UN Development Programme, UN Environment Programme, or the Food and Agriculture Organization to provide the needed expertise.

New technologies and techniques used in other countries related to effective land use can also be viewed as climate-related resources that can assist in the sustainable use of natural resources, including climate conditions.

One manifestation of the view that climate can be a resource is that societies have sought to modify the seasons to address their perceived needs, wants, and benefits, making them what might be called human-made seasons. Environmental modifications include irrigation, refrigeration, air conditioning, heating, use of food preservatives, dwelling design, and even the development of transportation infrastructure that enables the shipment of food stocks from locations where they are grown to locations where they are in demand but not grown. As a result, in a given region food availability (but not necessarily local access to it) is no longer dependent on the local climate or its seasonality.

Societies around the globe have taken advantage of their

climate conditions, even those located in what many people perceive to be harsh climate environments, such as the polar and hyper-arid regions. Where they failed to identify ways to cope with harsh climate or climate-related conditions, they avoided establishing settlements. At the same time, they figured out ways to exploit the natural environment on a seasonal basis. For example, over one hundred years ago, a U.S. Department of Agriculture yearbook (Newell, 1896) made reference to the awareness of land use in arid areas: "The success of agriculture in a distinctly arid region, like the valleys of Utah, where perennial streams flow from snow-capped peaks, is a self-evident proposition. There the climate renders irrigation absolutely essential and widely distributed. . . . No settler thinks for a moment of trying to cultivate the soil until he has provided a means of applying water" (p. 167). These are some of the positive, yet often unheralded, aspects of the climate system.

Each decade produces at least one attempt to understand how civilizations developed or collapsed. Diamond (1997) wrote a recent treatise about food production and diffusion. Throughout the book, he referred to the influence of climate on the development of sedentary agriculture and on the use of rangelands. He noted the importance of latitude, which is a determinant of climate conditions for the production of various kinds of food and livestock products. His writings challenge several historical viewpoints proposed in previous decades about where, when, and why agricultural societies were able to become established, as well as why societies sharing the same latitude and general climate aspects, such as length of day, failed to survive. His writings provide an interesting perspective on climate, not only as a resource but also as a constraint and a hazard to development.

When people think of climate's positive aspects, they tend to look at value added to the ways that they already exploit their climate regimes and use (or don't use) existing climate forecasts and other climate-related information. However, to the chagrin of climatologists and meteorologists, not every person and not every society is seeking to identify, let alone convert, potential climate knowledge into a value-added reality.

Climate as Hazard

Despite all of the positive aspects of climate for a region, governments for the most part are concerned about climate as a hazard. They want to avoid climate-related harm to their citizenry. However, they do not necessarily believe that it is their responsibility to enhance the value that their socioeconomic sectors generate from the climate system. They might support the efforts of others to reap profit through funding, subsidies, or demonstration activities, but they do not do so directly themselves. They leave this to private forecasters, the private sector, and entrepreneurs in climate-sensitive sectors. Although governments seldom get credit for pursuing additional benefits derived by using climate as a resource, they would certainly take the blame if a climate hazard became a disaster that causes death and destruction. Clearly, avoiding climate-related harm is much more important to their political well-being than is enhancing climate as a resource.

Although global warming and sea level rise are concerns of paramount importance to future as well as present generations, most of the world's population is more immediately concerned about getting through this season, this harvest, or this year. People have to figure out how to survive weather events and climate anomalies on these societally relevant but relatively short time scales. Basic issues related to energy and land use as they influence global warming are left to policy makers and scientists, who are paid by societies to deal with such monumental political, economic, and life-threatening concerns. The following paragraphs identify climate impacts of social importance. They are not listed in order of priority, other than referring first to climate phenomena and anomalies of major immediate concern—droughts and floods.

Droughts

Droughts occur somewhere on the globe every year. They have occurred throughout the history of humankind. Most people do not realize that there are different kinds of drought—meteorological, agricultural, and hydrological—and each one of which has its own impacts on society and environment. The most common perception of drought is a meteorological one, that is, a reduction in rainfall over a specified period of time by a certain percentage, e.g., 75 percent of average rainfall in the springtime or for the year. In a sense, this is an arbitrary determination. A precipitation reduction of 25 percent does not tell us much about its impacts on crops or water resources. For example, if the 75 percent that does fall comes during critical crop growth stages, then that reduction might not affect the plants or be of concern to the farmer. An agricultural drought can be said to have occurred if the rain that does fall, regardless of the amount, comes when it is not useful for the growth and development of field crops. To be sure, the farmer will take notice. Hydrological drought refers to a notable reduction in annual stream flow or water supplies for a given period of time.

Droughts often affect the quality as well as the quantity of food supply (now referred to as food security) within a country. Droughts do not respect international borders; the North American droughts of the 1930s affected Canada's prairie provinces as well as the United States Great Plains. Although many people have been led by their governments and the media to believe that droughts are the primary cause of famines, closer scrutiny of most famine situations uncovers a multistressed political environment, of which drought was but one factor. Depending on the specific situation, it is often not the most burdensome factor. A growing number of examples indicate that people have starved to death outside of grain storage facilities filled with grain. This situation occurred in the 1943 Bengal famine and in the 1970s famines in the West African Sahel and in Ethiopia (Sen, 1981). In an industrialized or a wealthy country, drought-related food shortages are often just an economic inconvenience, as prices for some commodities rise. However, the continued supply of those commodities is seldom in doubt. In a developing or poor society, the situation is quite different. Drought, especially an agricultural one, can easily combine with other stressful political or socioeconomic factors to increase the potential for famine.

Floods

Flooding is a highly visible aspect of variability in the climate system. It seems to capture the media's attention much more so than droughts. Although flooding is often a discrete event of relatively short duration, it can be highly devastating to life and property and extremely disruptive of human activities. Flooding also generates considerable misery among those directly affected by it because of the loss of irreplaceable mementos, temporary displacement, and even forced temporary or permanent relocation.

As with droughts, floods occur in many places on the globe every year. With the globalization of news coverage, people around the world with access to TV, radio, or newspapers are frequently exposed to stories about destructive and costly floods. Video coverage of the devastating Mozambique floods of 2000 captured the image of a woman giving birth in a treetop surrounded by raging floodwaters. That video led to flood disaster responses by governments that had been otherwise unresponsive. While writing these few paragraphs in mid-2002, floods were reported in the media in the United States (Texas and Minnesota), Venezuela, the Philippines, Bangladesh, eastern India, and southern Russia, among other locations.

Tropical Storms

Tropical storms (hurricanes, cyclones, and typhoons) are extremely destructive phenomena. They cause considerable damage to life and property around the globe due to the effects of strong winds, heavy rains, and tidal surges. Their frequency and intensity vary, producing adverse impacts wherever they make landfall and in the ocean as well. The Atlantic coast of the United States and other nations in and around the Caribbean Sea and Gulf of Mexico suffer to some extent almost every year during the hurricane season. Island nations and other countries in the South and Western Pacific, such as Australia (figure 2.1), the Philippines, China, and Vietnam, are adversely and positively affected by typhoons. Positive aspects include the bringing of

Figure 2.1. On 22 March 1999, Tropical Cyclone Vance, one of the strongest cyclones ever to affect mainland Australia, crossed the Pilbara coast of western Australia near the town of Exmouth. (Source: NOAA)

rainfall to otherwise water-short regions, such as the small island states in the Pacific. Many South Pacific islands depend on rainfall from tropical storms for their water supply.

The Bay of Bengal has seen some extremely destructive cyclones. Bangladesh has been particularly hard hit. The infamous 1970 cyclone caused upward of 300,000 deaths there (Ali, 1998). Some observers have suggested that the slow Pakistani governmental response to the cyclone damage in the eastern part of that country provided yet another reason for the inhabitants there, separated by India from the western part of Pakistan, to demand independence for Bangladesh (Lewis, 1999).

India's state of Orissa, one of its poorest regions, suffered major losses (more than 20,000 dead, about $4.5 billion of destruction) as a result of the 29 October 1999 supercyclone. The misery factor was also very high: 15 million homeless or displaced persons, rice crops destroyed, hundreds of thousands of cattle deaths, salinization of coastal soils, a high rate of joblessness, and reduced forest cover (UNICEF, 2000).

There is concern in scientific and policymaking circles about how global warming might affect the frequency, intensity, trajectory, and location of landfall of tropical storms. For example, in the 1990s alone, devastating hurricanes such as Andrew and Floyd in the United States, Georges in the Caribbean, and Mitch in Central America have heightened concern about the possibility of increased damage from blockbuster hurricanes. Landfall of a storm of the same intensity and in the same location in the United States would have a much greater economic impact today than it would have had fifty years ago because there has been a major growth of population along U.S. coastal areas. Researchers have sought to model the possible impacts of very intense hurricanes, called "supercanes" (Emanuel et al., 1995).

Researchers have noted a fluctuation of a few decades in the frequency of tropical storms in the Atlantic-Caribbean region, as suggested by figure 2.2. This makes it difficult, at least in the short term, to attribute an increase in more tropical storms either to global warming (deep climate change) or to a return to an earlier period when there were more storms (fluctuation).

Ice Storms and Frosts

Severe ice storms are not as frequent, or at least not as noteworthy, as other weather extremes. However, they do occur locally every winter in the Northern Hemisphere. They are often highly disruptive of energy, communications, and transportation infrastructure. A notable storm, called the Ice Storm of the Century, occurred in the central United States in 1993 (Sharn and Smaragdis, 1993). Another blockbuster storm

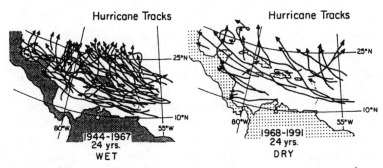

Figure 2.2. Hurricane track density 1944–1991. (Source: Gray et al., 1997)

occurred in Canada in January 1998. Labeled one of Canada's worst ice storms in history, it destroyed power lines, closed businesses, disrupted commerce, damaged infrastructure, and deprived thousands of people of heat for several weeks in the midst of a Canadian winter (Environment Canada, 1998).

Frosts and freezes can also be costly to weather-sensitive industries and activities. An untimely frost can have a major adverse effect on agricultural production. For example, in the 1980s, an unexpected succession of frosts and hard freezes in central Florida had major adverse impacts on orange groves and in turn on the American frozen concentrate of orange juice (FCOJ) industry in that state. In response to freezes in Florida, Brazilian entrepreneurs made a strategic marketing decision to export large quantities of FCOJ to the lucrative United States orange juice marketplace. This developed into a strong export-oriented industry in that country and, therefore, a major source of foreign exchange for Brazil (Miller and Glantz, 1988; see also McPhee, 1967, or the Hollywood movie *Trading Places*).

In a developing country, a severe frost can create a life or death situation for those dependent on subsistence agriculture. Drought-induced frosts occurred in the Papua New Guinea highlands during the 1982–1983 El Niño (Allen, 1993) and again during the 1997–1998 El Niño and led to severe food shortages and contributed to famine in the country (Barr, 2001). The 1997 famine was the only one to have occurred during that El Niño.

Fires

Throughout the 1990s, wildfires plagued several countries—the United States, Mexico, Australia, Russia, and Indonesia, among others. Fires destroy the natural and human-built environments in both rural and urban areas. The connection between climate conditions and wildfires is easy to understand. During periods of extreme meteorological drought and persistent high temperatures, forests and grasslands are susceptible to lightning strikes, which are frequent catalysts for the large-scale burning of dry underbrush and grasses. In many instances, however, fires have been started by human activities, purposely as well as accidentally. Major fires in Colorado and in Arizona in the summer of 2002 were purposely set. Extremely dry conditions provided the underlying cause for fire and the arsonists provided the catalyst—matches.

According to Glover and Jessup (1999),

> The year 1997 was the worst on record for forest and bush fires throughout the world. . . . It was, in the words of the World Wide Fund for Nature (WWF), 'the year the world caught fire.' Catastrophic fires occurred in Indonesia, Brazil, and other countries across Asia and the Pacific, Latin America, and Africa (p. 1).

Arguably one of the twentieth century's most notable fires took place in Indonesia during the 1997–1998 El Niño, when an estimated 9 million hectares of forestland burned. Since the 1982–1983 fires in Indonesia, when about 3 million hectares of rainforest burned, people have identified a strong connection between El Niño events and rampant Indonesian fires in Kalimantan on the island of Borneo. Under normal agricultural conditions in the country, fires are set each year to clear the postharvest crop residue in the cultivated fields (figure 2.3). Farmers do so because they believe that such fires put nutrients back into the soil, and it is an easy and useful way to clear the land for cultivation. Thus, many of the fires that spread out of control in Kalimantan and on Sumatra during the 1997–1998 El Niño had been intentionally set. Although the occurrence of widespread fires was at first blamed on El Niño, investigations showed that

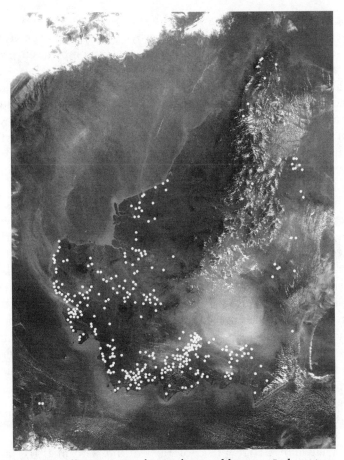

Figure 2.3. Satellite picture of 2002 fires and haze on Indonesian island of Borneo. The white dots represent fire locations. (Source: Visible Earth, 2002; taken by Jacques Descloitres, MODIS Land Rapid Response Team, NASA/GSFC, 19 August 2002)

many of the fires were not the result of traditional land-clearing practices. Agricultural company executives in collusion with governmental officials seeking to circumvent national laws that were designed, in theory at least, to protect the country's remaining forested areas had illegally and clandestinely commissioned many of the fires to be set, knowing that an El Niño event would bring drought to the region. According to the law, forests in Indonesia that happened by chance to be destroyed as a result of natural causes, such as lightning, could then be used by companies to grow lucrative plantation crops, such as palm oil.

These fires produced thick haze throughout the region, affecting the air quality not only in Indonesia, but also in various countries of Southeast Asia. The regional extent of haze was captured by satellite, making it impossible for Indonesian authorities to deny responsibility. Two passenger plane crashes, two ship collisions, and various regional human health problems (e.g., respiratory ailments) were blamed on the haze.

Industrialized countries are also at risk to wildfires. Deadly fires, referred to as the Ash Wednesday Fires, occurred in towns around Melbourne, Australia, during the 1982–1983 El Niño. Lives were lost, buildings were destroyed, and almost 400,000 hectares of land surface burned (Oliver et al., 1984). The Black Christmas Fires began on Christmas Eve 2001 in Sydney, Australia and ran into early January 2002. Many of these fires were maliciously set by people and were not the direct result of lightning (Halperin, 2002). Drought-related forest fires and their adverse impacts on flora, fauna, and human settlements have also plagued relatively large areas of the western United States, western Canada, and Russia's Far East regions.

Infectious Diseases

Malaria, dengue fever, cholera, equine encephalitis, and hantavirus, among other infectious diseases, are associated with climate conditions. The climate-disease interaction has recently become a focus of interest for governmental, nongovernmental, and humanitarian funding agencies, and for health and climate impacts researchers. That interest has been catalyzed by unexpected climate-related disease outbreaks such as the cholera pandemic in Latin America, which was linked to the 1991–1992 El Niño. Cholera is still a major threat in Latin America and other impoverished regions of the world. Figure 2.4 traces the Latin American epidemic that began in Peru in 1991 (Hoff and Smith, 2000, p. 21).

El Niño events have also been associated with dengue fever

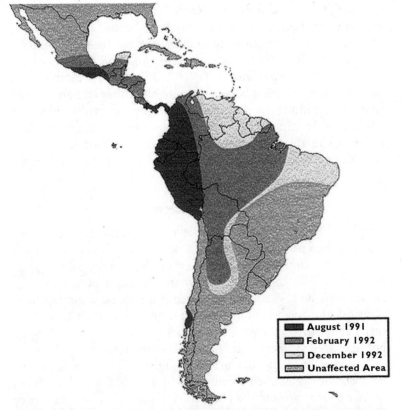

August 1991
February 1992
December 1992
Unaffected Area

Figure 2.4. Spread of Latin American cholera pandemic. Reprinted with permission. (Source: Hoff and Smith, 2000, p. 21)

outbreaks in Vietnam, malaria outbreaks in Venezuela and Colombia, and a Rift Valley fever outbreak in Kenya during heavy flooding in that country in 1997–1998.

Perhaps the primary catalyst for growing concern in the 1990s about climate and infectious disease patterns and outbreaks centers on global warming. If global warming continues, tropical diseases are likely to spread poleward, bringing tropical disease vectors such as mosquitoes to the temperate climate zones of the midlatitudes. Signs of this trend are already appearing. This nascent concern has led to a sharp increase in research on the climate-health nexus. Before 1990, the issue of climate's influence on health was left on the back burner by the climate change research community and by climate research funding agencies. Health was the responsibility of the United Nation's World Health Organization (WHO, 1996). For example, in a paired comparison test between health and other climate-related issues, the funding of climate-health issues seemed to place far behind the funding of climate impacts on agriculture, water, and energy. The relative positions have since shifted in favor of human health and climate interactions (Colwell and Patz, 1998).

Urban Impacts

People in both industrialized and developing countries are migrating to urban areas in search of a better life. Governments are increasingly concerned about whether their metropolitan areas can continue to cope with rapidly increasing populations (e.g., Rosenzweig and Solecki, 1999). Aside from the need for expanded physical and socioeconomic infrastructure and aside from concerns about urban air pollution or land annexation, an emerging but neglected problem in urban areas is urban water supply.

It is highly likely that urban drought will become a prominent climate-related environmental and human health problem in many areas around the globe during the twenty-first century (Changnon, 1994). The list is growing of major cities that have already suffered from urban drought, even in industrialized countries—New York City, Los Angeles, Chicago, Tucson, Washington, D.C. Such water shortages cause water quality problems as well. Urban drought, or more broadly, urban water shortage, is not just a concern of the developed countries but even more so of developing countries. In his book, *The Future in Plain Sight*, Linden (1998) suggested that

> In the poorer nations of the world, the latter part of this century has seen a massive, unprecedented migration to the cities. . . . Urban migration offers a possible soft landing for these mounting pressures in the developing world. On the other hand, it could just as easily spur social, political, epidemiological, and environmental

collapse. . . . Sometime in the next [twenty-first] century, cities may have to deal with the consequences of climate change. Even if this threat proves to be a chimera, population pressures have raised the stakes of ordinary climate variations. . . . [In addition], should the world see more frequent and intense storms in coming years, hundreds of millions more stand to suffer, since thirty of the world's fifty largest cities lie near coasts (pp. 52, 60–61).

Thus, in addition to the influence of climate on the quantity of water flowing to urban areas, other demographic and industrialization processes are simultaneously at play. This combination of natural and societal factors creates severe water problems in urban areas.

Population Migration and Climate

Most population-related issues are highly provocative topics in international political circles. One key climate-related concern is population movement as populations increase. Because most of the best rain-fed agricultural land is already in production, people are increasingly forced to move into areas that are marginal for agricultural activities—marginal in terms of rainfall, water availability, soil quality, topographic features, or altitude. When farmers migrate into new areas, they often rely on the same cultivation practices or rangeland stocking rates that they used in the relatively less marginal areas from which they came. However, environmental conditions in the newly inhabited marginal areas are often much less favorable. Thus, more crop failures are likely to occur and are likely to be blamed on climate, rather than on land-use practices that may be ill-suited to the local environmental conditions. This process has been referred to as "drought follows the plow" (Glantz, 1994a).

Recreation and Ecotourism

Weather and climate are not always a life, death, or economic concern. People are also concerned about the weather and climate for reasons related to recreation. Ecotourism has become a thriving industry. People plan vacations based on what they believe the atmospheric conditions are likely to be at their destinations. For example, those who like to sail on lakes or swim in the ocean pay close attention to the marine meteorological forecasts (tropical storms, winds, wave heights, storm surges, etc.). Those who fail to pay close attention to such forecasts do so at their own peril (Mundle, 2000).

Owners of resorts select the locations for their facilities based on expected climate and extreme weather conditions as much as on other factors. For example, ski resorts need natural snow conditions to lure prospective skiers and when it is late in coming or lacking in amount, they often make snow artificially. Cloud-seeding activities in mountainous areas are attempts both to improve skiing conditions in winter and increase water supplies for use in summer.

Financial Aspects

The financial world also has legitimate climate concerns. Insurance corporations have taken a keen interest in climate variability, change, and extremes as a direct result of the cluster of billion-dollar payouts for climate- and weather-related disasters in the 1990s. These losses prompted stockbrokers to begin a trade in weather derivatives. The Chicago Mercantile Exchange (CME) was the first to engage in weather derivatives transactions involving futures and options on futures related to an index of so-called heating degree days (HDD) and cooling degree days (CDD).* Degree days refers to the changes over a month or a season (calculated on a daily basis) from the long-term average of the temperature in a given location, using, for example, 65°F as the baseline temperature for the continental United States.

Derivatives represent one aspect of a broader and increasingly more active weather risk market. Weather derivative futures are "legally binding agreements to buy or sell the value of the HDD/CDD index at a specific future date."

---

*How HDD and CDD indexes are calculated can be found at the following website: (www.cme.com/products/index/weather/products_index_weather_about.cfm).

Companies whose activities are directly affected by weather or climate variability (for example, for heating or for cooling) can engage in a risk management approach by trading in temperature-related weather derivatives. According to the CME, "Just as professionals regularly use futures and options to hedge their risk in interest rates, equities and foreign exchange, now there are tools available for the management of risk from extreme movements of temperature." This has also been called the "securitization of weather risks."

Innumerable scenarios could be constructed related to unanticipated weather or climate variations that can positively or negatively affect commerce, trade, recreation, and tourism. A relatively cool summer can reduce energy bills because of a reduced need for air conditioning but could also reduce visits to beach resorts. A cold snowy winter could cause a sharp increase in energy demand. An El Niño-related warm winter would help to reduce energy needs and demand in the northeast United States, and so forth.

The *Japan Times* recently (22 November 2001) reported that two Japanese corporations had entered an agreement to cover each other's losses that might be incurred as a result of seasonal variability. Specifically, the Kansai Electric Power Co. (Kepco) and the Osaka Gas Co. agreed to the following: "When Kepco generates greater profits than expected in a scorching summer due to increased power consumption for air conditioners, it will pass on part of the profits to Osaka Gas, which usually sees lower gas demand when temperatures are high. The two companies will do the opposite in cool summers that undermine the power firm's profits while boosting those at the gas company."

CME noted that "weather, universally unpredictable, can destroy profits." If it's true that the weather is unpredictable, then weather derivatives might just give weather-sensitive sectors of society a fighting chance to improve their odds toward neutralizing the financial losses of weather variability and extremes that affect their operations.

For the first time, big companies have gotten together without the assistance or involvement of government to engage voluntarily in a carbon trading activity, called CCX (Chicago Climate Exchange). The idea was fostered by the belief that, despite the current US rejection of the Kyoto Protocol, eventually there will be controls placed by governments on carbon emissions. Twenty-eight companies signed up to reduce carbon emissions by a relatively small percentage at first and to trade carbon credits when their emissions fall below the agreed-upon reductions. The CCX was set up at the end of 2002, and the companies are grappling with the same issues as are governments, for example, how to calculate baseline emissions; how to swap (offset) emissions in one location by assuring emission reductions in other locations, such as through reforestation; how to get credit for carbon sinks (CCX, 2002). It is an experiment, but whether it works is less important than the fact that businesses are taking greenhouse gas emissions more seriously than ever.

Impacts at the Global Scale

The following three climate examples—El Niño-La Niña, global warming, and sea level rise—refer to the climate anomalies themselves and the impacts on physical systems of changes in atmospheric conditions. Each of these physical changes in turn sparks changes in socioeconomic and political systems in various locations around the globe.

El Niño and La Niña

The 1997–1998 El Niño has been called the El Niño of the twentieth century, replacing the 1982–1983 event that had previously held that title. Researchers and forecasters were truly surprised to have witnessed two such major events in fifteen years. They were surprised again in May 1998 when the El Niño collapsed as quickly as it had developed. Media coverage of the event helped to educate the public worldwide about the phenomenon and its potentially destructive capabilities. When people now hear that an El Niño has been forecast, they are most likely going to think back to their country's climate situation in 1997–1998 to forecast by analogy its possible impacts. In many places that experience was an unpleasant one, as lives were lost and properties destroyed.

By early 1998, forecasts of an emerging La Niña episode

began to appear. In some locations, La Niña has been associated with climate anomalies opposite to those that occur during El Niño. For example, the Philippines, Indonesia, Vietnam, and Australia, among others that are affected by drought during El Niño, are at an increased risk of floods during La Niña. As with any climate-related extreme, the severity of impacts depends on a combination of the intensity of La Niña and the level of societal vulnerability.

The tropical Pacific is the field of action for the El Niño Southern Oscillation (ENSO) cycle, and as a result, the countries around the Pacific Rim appear to be among the most directly affected by the cycle's warm and cold extremes. How global warming might affect the behavior and characteristics of the ENSO cycle and its impacts on ecosystems and societies has generated considerable scientific debate and research. However, the scientific underpinnings of the ENSO-global warming relationship have not as yet been uncovered (Greenpeace, 1997).

Global Warming

Since 1975, international interest in the human (anthropogenic) contributions of greenhouse gases and land-use changes to, and impacts on, the naturally occurring greenhouse effect has grown markedly. Since then, scientists have been brought together on numerous occasions to assess the state of scientific understanding of greenhouse gases and global warming. Ever since the late 1980s, government representatives have been discussing and then negotiating approaches to reduce national contributions to global warming by reducing greenhouse gas emissions. Many governments signed the United Nations Framework Convention on Climate Change (UNFCCC) at the Earth Summit of 1992, including the United States, the largest emitter of CO_2. Internationally, governments have since convened eight meetings as of 2002 for negotiations, referred to as the Conference of Parties (COPs). Considerable funding has supported attempts to model and understand atmospheric processes, the greenhouse effect, and how human activities and ecological processes might be altering and be altered by them.

Because sharp increases in fossil fuel burning and resulting carbon dioxide emissions are at the center of the global warming debate, and because economic progress in the industrialized countries has been based on the use of such fuels, concerns about global warming have devolved into a debate about energy policy and a search for environmentally friendly approaches to sustainable economic development.

Sea Level Rise

Climate change concerns center on two key changes in the physical environment: global warming and its impacts on regional temperature and precipitation, and a resultant rise in global sea level. Sea level rise could result primarily from a combination of the melting of glaciers, Antarctic and Arctic ice sheets, and a thermal expansion of the oceans. Increases in sea level threaten low-lying islands and coastal areas worldwide (figure 2.5). More than a dozen of the world's major cities are located along coasts or estuaries and are equally threatened by sea level rise. One of the most active organizations at the intergovernmental level to focus attention on the sea level rise issue has been the Association of Small Island States (AOSIS). AOSIS actively lobbies for the reduction of greenhouse gas emissions by industrialized countries.

Bangladesh, among the world's poorest nations, is one of the countries most vulnerable to sea level rise. Upward of 20 million inhabitants of the country are already at risk to coastal storm surges that result under present-day sea level conditions. Catastrophic storm surges in the past have caused damage up to 100 km inland. Any rise in sea level, natural or human-induced, greatly increases the country's vulnerability.

The Netherlands' engineering expertise for coastline protection techniques has been in demand by areas at risk to rising sea level (e.g., Louisiana and Bangladesh). Through its levees and sea walls, the country's engineers have successfully protected the population from floods over the centuries, with at least one notable exception in recent decades (in 1953). The flooding that occurred in the country in 1993 and again in 1995 resulted from decades of engineering attempts to control and channel the flow of major rivers such as the Rhine.

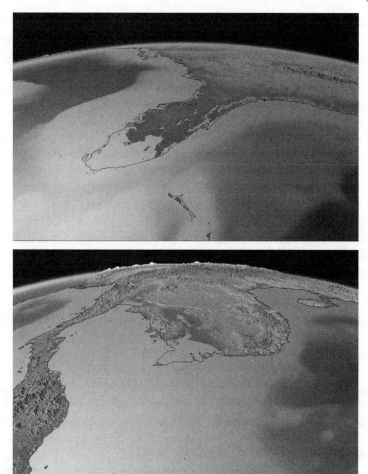

Figure 2.5. Projected impacts of sea level rise in Florida, USA and Southeast Asia with a 5-meter sea level rise. (Reprinted with special permission of William Haxby, Columbia University)

Summary

In sum, climate anomalies are occurring somewhere on the globe every year. Some of those anomalies are induced by global-scale changes in the atmosphere, while others are either regionally or locally induced. The end result that their impacts might have on society is determined by the combination of influences from all of these levels. Then there are the influences of human activities in situ that help to determine the severity of the climate-related anomaly. This cauldron of factors affecting climate-society-environment interactions at any given place and time is also affected by the fact that today's climate warming is consistent with a deep climate change. How global warming will influence each of these factors and their interactions is something we cannot know with confidence, thereby making reliable forecasts about those interactions difficult at best.

As this chapter is being written, news about a whole new set of anomalies is appearing in the headlines: floods in northern India, China, central Europe, and West Africa; droughts in India, Zimbabwe, Malawi, Australia, and the United States; freezing temperatures in some South American countries; fires in Indonesia; and talk of El Niño's return. However, this set of anomalies does not exactly match any other annual set. Therefore, it is very important that decision makers stay alert to the prospects of damaging climate anomalies and encourage the development of early warning systems as well as a wide range of effective coping strategies to draw upon in the event of climate-related emergency situations. By analogy, the appropriate time to fix a potentially leaky roof is before the rainy season starts and not during a rainstorm.

The next chapter provides a glimpse of the anomalies that currently plague settlements on various continents. It is meant merely to suggest the types of extremes that can occur and the levels of vulnerability to climate anomalies that those settlements might have. To date, there is relatively little sharing of climate-related experiences across continents, even though people and governments may have to face the same types of climate-related hazards.

THREE

CLIMATE AND GEOGRAPHY

No part of the earth is immune from variations within the global climate system. Inhabitants of each continent, region, and sub-region have their own set of climate concerns and climate-related hazards to worry about. It is not possible to capture all of the major climate anomalies that can or do affect a particular continent in just a few paragraphs. Yet, a few examples of recent climate, its anomalies, and impacts on the various continents show the reality that no part of the globe is immune to climate variability, fluctuations, and extremes. As a relatively more detailed example of the many ways that climate can influence the environment and the inhabitants of a region, this chapter concludes with a case example of climate and African development.

As time progresses, societies tend to downplay or forget past struggles with climate. Instead, they tend to play up their attempts to predict the onset of climate extremes, to manage the impacts of those extremes, and to identify the physical causes of climate anomalies that can affect their economies and the well-being of their citizens. Most encyclopedias, in describing the climatic regime of a given country or region, give the impression that the climate is stable, that climate characteristics are constant. As we now know, however, climate is defined by ranges and frequencies of behavior that are anything but consistent and predictable. Although it is rarely done, societies need to make the effort to expose each new generation to the ways that people in the region were affected by climate anomalies in the past, as well as the ways they might be affected by climate variations of the future. Creating climate awareness is not a one-time event. It is a lifelong, ongoing process.

The following paragraphs provide a reminder of the kinds of climate and weather events that occur each year in the different regions of the globe. Most people, including those deeply immersed in the information age and constantly traveling the Internet, are unaware that these anomalies or their impacts ever happened. But they did, and they were costly, if not to an insurance company, then to the communities affected by them. Table 3.1 was compiled by Munich Re (2000), a major global reinsurance corporation that is extremely concerned about climate change and climate extremes.

Researchers at Munich Re believe that there has been, and will continue to be, an increase in what they consider natural catastrophes. They cite as the reasons for such an increase population growth rates, rising standards of living, the concentration of populations and property values in metropolitan areas, settlement and industrialization of very exposed regions, vulnerability of modern societies and technologies, and changes in environmental conditions. Munich Re's researchers also believe that global warming will increase the frequency and intensity of extreme events and will affect the spatial range of their impacts.

Figure 3.1 provides a glimpse of global meteorological impacts in a given year, 1999. In a sense, it was an unusual year because it was a full calendar year embedded within the lengthy 1998–2001 La Niña episode of anomalously low sea surface temperatures in the central and eastern equatorial Pacific.

North America

Droughts, floods, severe storms, hurricanes, freezes, blizzards, and fires are responsible for many of the continent's notable climate-related problems (see box 3.1 for what is considered normal).

Table 3.1.
Significant Natural Disasters in 1999

Date	Region	Event	Deaths	Economic Losses ($USm)	Insured Losses ($USm)
Jan–March	Switzerland, Austria, Germany	Avalanches	100	600	50
Jan. 1–4	USA, Canada	Winter storm, tornadoes	18	1000	755
Jan. 13–16	USA, Canada	Winter storm, ice storm	0	755	575
Jan. 25	Colombia	Earthquake	1185	1500	150
Apr. 14	Australia	Hailstorm	1	1500	1000
May 3–7	USA	Tornadoes	51	2000	1485
May 20–23	Pakistan, India	Cyclone 02A	644	20	
May 12–28	Germany, Switzerland	Floods	10	850	300
June–July	China	Floods	800	8000	
Aug. 17	Turkey	Earthquake	17,200	12,000	1000
Sept. 7	Greece	Earthquake	138	4150	150
Sept. 14–16	USA, Bahamas	Hurricane Floyd	61	4000	1800
Sept. 20	Taiwan	Earthquake	2470	14,000	850
Sept. 21–Oct. 15	Mexico	Floods, landslides	500	230	
Sept. 22–25	Japan	Typhoon Bart	26	5000	3000
Oct. 29–Nov. 11	India	Cyclone 05B	15,000*	2500*	115
Dec. 3–4	Western and Northern Europe	Winter storm Anatol	20	800	400
Dec. 13–16	Venezuela	Floods, landslides	20,000*	15,000*	500*
Dec. 26	Central Europe, France, Switzerland, Germany	Winter storm Lothar	100*	7500*	4000*
Dec. 27	France, Spain, Switzerland	Winter Storm Martin	30*	2000*	1000*

*indicates estimates.
Source: Munich Re (2000)

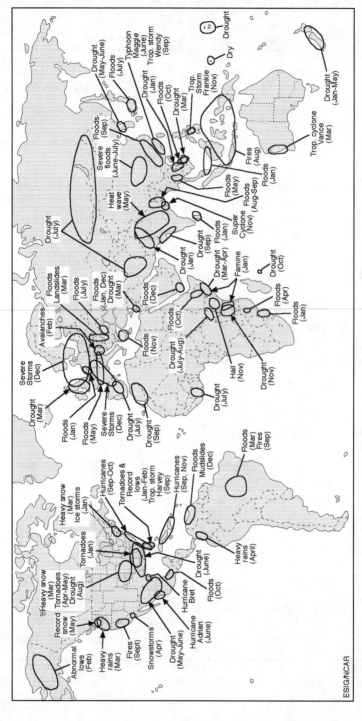

Figure 3.1. Climate anomalies in 1999, a La Niña year. Not all anomalies that occurred during 1999 can be attributed to La Niña. A La Niña cold event developed in March 1998 and extended into early 2000. (Source: Glantz, 2001a, p. 82)

Given the early warnings for several of these potential climate and weather-related hazards, the death toll attributed to them in North America sharply declined over the 1900s. On occasion, however, a major disaster occurs during which a relatively large number of people die—the 1999 heat wave in Chicago, Illinois, for example. This high-impact weather event proved once again that the poor and the elderly are often the primary victims of excessive heat.

Heat waves raise an interesting point. Is a heat wave a weather extreme or a climate extreme? The Chicago example just cited suggests that heat waves are short-term (days) meteorological events that can lead to an increase in the number of heat-related deaths. Yet another way to measure the impacts of heat, however, is to look at entire seasons during which temperatures are well above a normal range. The *Time Almanac* (Brunner, 2001, p. 625) listed some of the twentieth century's major heat-related disasters. The 1980 and 1999 heat waves were cited as climate-related and the 1995 situation as a weather event:

> *1980 Summer*, central and eastern US: a severe drought and heat wave killed an estimated 5,000–10,000 people, including heat stress-related deaths.

> *1995 July 12–17*, US Midwest and Northeast: over 800 persons, including 560 in Chicago, died in a record heat wave.

> *1999 Summer*, continental US: record heat continued throughout the country, resulting in drought, crop damage, and 282 deaths nationwide.

Each part of the United States has its own hazards throughout the seasons: tornadoes in the Midwest, droughts in the High Plains and Northwest, floods in various locations, forest fires in the West, blizzards and ice storms in the Northeast, hurricanes in the Gulf Coast, Southeast, and mid-Atlantic regions. It is very difficult to find a single weather or climate calamity with which the entire nation has to cope, because of its varying climate zones (box 3.1).

Often overlooked aspects of climate impacts in North America are the opportunities created for other countries to take advantage of North America's climate problems. Two Brazilian examples come to mind: frozen concentrate of orange juice (FCOJ) to the United States and soybean exports. Brazilian investors first

BOX 3.1
Climate Zones of North America

North America possesses a multitude of diverse regional climates as a consequence of its vastness, its topography, and being surrounded by oceans and seas with widely varying thermal characteristics. The North American region extends from approximately the Arctic Circle to the Tropic of Cancer and from the Aleutian Islands in the west to the Canadian maritime provinces in the east.

In the colder half of the year, the position of the Polar Front, which generally separates colder, drier air to the north from warmer, moister air to the south, can vary greatly from southern Canada to the southern United States. Large shifts in the Polar Front are associated with changes in the atmosphere that often cause one part of the continent to experience warm, moist, southerly airflow while another part experiences a blast of dry, cold Arctic air. However mainly in the fall and spring, shorter, weaker waves in the atmosphere move more quickly across the continent—producing highly variable weather with rapidly changing, but not extremely high or low, temperatures and short wet and dry periods. In the summer, the Polar Front retreats well into Canada, for the most part, and two oceanic semi-permanent high-pressure systems tend to dominate the North American weather.

The temperature regime over North America varies greatly. Over all seasons, mean temperatures generally increase from the extreme north along the Arctic Ocean to the southern United States. Mean annual and wintertime temperatures along the west coast of the continent generally are higher than at equivalent latitudes inland or on the east coast because of the warming influence of Pacific air. West of the Rockies, warmer maritime airflow off the Pacific Ocean produces milder winters along the coast; the western cordillera effectively restricts this mild air from reaching and thus modifying temperatures in the interior. The eastern regions of the continent enjoy much less warming influence from the Atlantic Ocean during these cold air outbreaks because the prevailing air flow is off the land. Nevertheless, in winter the east and west coastal regions of Canada and the United States usually are warmer than inland regions, with the Pacific and Gulf coasts and Florida experiencing the shortest and mildest winters.

In summertime, the large amount of solar radiation received over very long days in the northern reaches of North America acts to raise

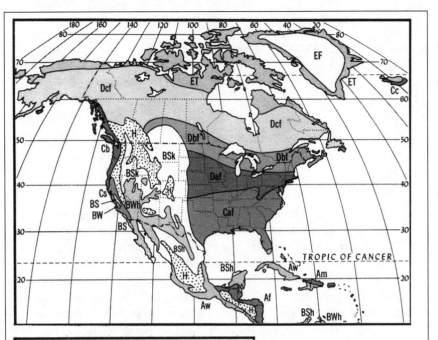

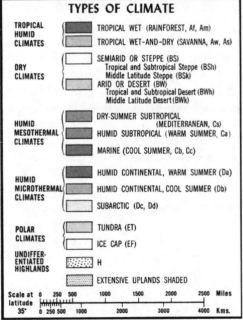

TYPES OF CLIMATE

TROPICAL HUMID CLIMATES
TROPICAL WET (RAINFOREST, Af, Am)
TROPICAL WET-AND-DRY (SAVANNA, Aw, As)

DRY CLIMATES
SEMIARID OR STEPPE (BS)
  Tropical and Subtropical Steppe (BSh)
  Middle Latitude Steppe (BSk)
ARID OR DESERT (BW)
  Tropical and Subtropical Desert (BWh)
  Middle Latitude Desert (BWk)

HUMID MESOTHERMAL CLIMATES
DRY-SUMMER SUBTROPICAL (MEDITERRANEAN, Cs)
HUMID SUBTROPICAL (WARM SUMMER, Ca)
MARINE (COOL SUMMER, Cb, Cc)

HUMID MICROTHERMAL CLIMATES
HUMID CONTINENTAL, WARM SUMMER (Da)
HUMID CONTINENTAL, COOL SUMMER (Db)
SUBARCTIC (Dc, Dd)

POLAR CLIMATES
TUNDRA (ET)
ICE CAP (EF)

UNDIFFERENTIATED HIGHLANDS
H
EXTENSIVE UPLANDS SHADED

Scale at latitude 35°
0 250 500 1000 1500 2000 2500 Miles
0 250 500 1000 2000 3000 4000 Kms.

Source: Trewartha, 1961

temperatures there so that these areas are more in line with much of the rest of the continent.

A main continental maximum in annual precipitation is in the southeastern United States. Another precipitation maximum is typically seen over the U.S. Midwest in the summer months, where mean rainfall typically exceeds 25 cm.

Source: www.grida.no (Global Resource Information Database Arendal)

entered the FCOJ export business in 1962 in response to freezes in Florida's orange groves. A major boost to the volume of their FCOJ exports to the United States came as a result of a cluster of yearly freezes in Florida in the early 1980s.

Brazil's decision to enter into soybean production was a reaction to the 1972–1973 collapse of the anchovy fishery in Peru, which was brought on in part by El Niño. That fishery was the source of the raw material for the production and export of fishmeal, which is used as an animal feed supplement for the broiler chicken industry in North America. When fishmeal became too expensive because of a sharp reduction in supply, the broiler industry in the United States resorted to using soy meal as a feed supplement. Today, Brazil is a leading global exporter of soybeans. Weather- and climate-related extremes sparked the development of this and the FCOJ export industries. This suggests that it pays for entrepreneurs to keep track of the climate anomalies not only in their own countries but in other countries as well. The United States' northern neighbor is also affected by seasonal-to-interannual climate variability, fluctuations, and change.

The way that foreigners view Canada, and even the way many Canadians see themselves, has been shaped for the most part by the country's perceived relatively harsh climate. According to Environment Canada's website, "Canada's climate often forms a big part of the perceptions that non-Canadians have when they think of Canada. When thinking of Canada, people of many other nations think of a country full of big, cold, snowy expanses. In reality, we are a land of distinctive and diverse climatic regions." The perceptions of a harsh climate were reinforced for decades by the way that Hollywood portrayed the country in feature films. Environment Canada challenges this perception by noting that "We are a land of seasons: spring, summer, fall, and winter flowing in a natural rhythm. The weather of our seasons can vary dramatically from region to region" (Environment Canada, 2002).

As with the United States, Canada's continental expanse and varied topography ensures that the likelihood of a national climate-related hazard is small but not zero.

Europe

The climate zones of Europe are varied, with some parts of the continent quite wet and others prone to aridity (see box 3.2).

However, as in most places, average conditions do not adequately describe a region's climate regime, but they provide a good place to start. Often, there are wet and dry extremes in various locations in a given year somewhere on the continent.

Europe, from its westernmost point on the coast of Portugal to Russia's Ural Mountains, faces a variety of anomalous climate episodes in all four seasons. Winter storms, heavy snows, strong winds, droughts, and floods plague various parts of Europe. For example, all of the following extremes, among others, occurred during the 1990s: hurricane-force winds struck England; strong winds destroyed tens of thousands of trees throughout Paris; extremely cold winters adversely affected the European part of the Russian Federation; floods impacted parts of Poland, Germany, and The Netherlands; and record-setting heat waves occurred in various cities throughout the continent. In mid-2002, central European floods proved to be among the most costly in the past one hundred years in eastern Germany and Czechoslovakia.

In mid-2002 at the World Summit on Sustainable Development in Johannesburg, South Africa the European Union ratified the Kyoto Protocol, which is an international agreement to reduce to 1990 levels greenhouse gas emissions within the next decade. It has apparently taken over the front-runner position from the United States on the global warming issue. European researchers are actively engaged in generating and evaluating computer-derived scenarios for a warmer earth and the implications for various economic and social sectors in European countries.

BOX 3.2
Climate Zones of Europe

Europe's particular distribution of land and sea—which includes several major inland seas such as the Mediterranean, the Baltic, and the Black Sea—and its long coastline facing the eastern North Atlantic Ocean are factors that help to shape the continent's numerous regional climates. The presence of various high mountains, which act as physical barriers to atmospheric flows, is responsible for substantial regional differences in precipitation patterns.

Although much of Europe lies in the northern latitudes, the relatively warm seas that border the continent give most of central and western Europe a temperate climate, with mild winters and summers. The prevailing westerly winds, warmed in part by their passage over the North Atlantic Ocean currents (the Gulf Stream), bring precipitation throughout most of the year.

In the Mediterranean area (i.e., Spain, southern France, Italy, southern Croatia, Montenegro, Macedonia, Albania, and Greece), the summer months usually are hot and dry; almost all rainfall in this area occurs in winter. From central Poland eastward, the moderating effects of the seas are reduced; consequently, drier conditions prevail, accompanied by a greater amplitude of annual temperature variations (i.e., hot summers and cold winters). Northwestern Europe is characterized by relatively mild winters, with abundant precipitation along the Scottish and Norwegian coasts and mountains, and much colder winters and generally drier conditions in Sweden and Finland. In mountain regions such as the Alps, winters are generally cool, and snow remains on the ground for several months of the year; summers are typically cool and moist.

The variety of European climates is determined by latitude or altitude as well as by proximity to the ocean or to one of several inland seas. Annual temperature ranges vary from some 10°C in coastal regions of the United Kingdom and Ireland to about 30°C in Finland and Russia. Annual precipitation totals range from as low as 200 mm in southern Spain and Greece to more than 2000 mm in coastal regions of Scotland and Norway and at some locations in the Alps.

Source: www.grida.no (Global Resource Information Database Arendal)

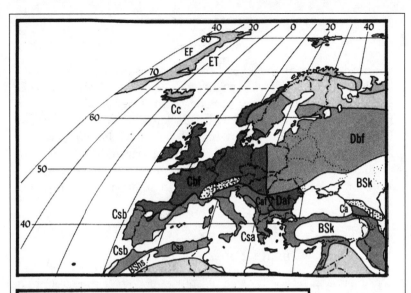

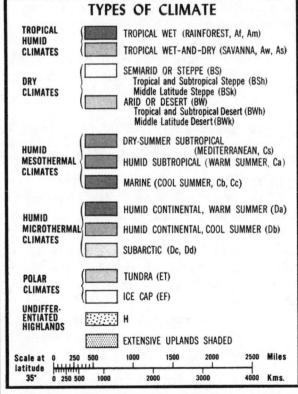

TYPES OF CLIMATE

TROPICAL HUMID CLIMATES
- TROPICAL WET (RAINFOREST, Af, Am)
- TROPICAL WET-AND-DRY (SAVANNA, Aw, As)

DRY CLIMATES
- SEMIARID OR STEPPE (BS)
  - Tropical and Subtropical Steppe (BSh)
  - Middle Latitude Steppe (BSk)
- ARID OR DESERT (BW)
  - Tropical and Subtropical Desert (BWh)
  - Middle Latitude Desert (BWk)

HUMID MESOTHERMAL CLIMATES
- DRY-SUMMER SUBTROPICAL (MEDITERRANEAN, Cs)
- HUMID SUBTROPICAL (WARM SUMMER, Ca)
- MARINE (COOL SUMMER, Cb, Cc)

HUMID MICROTHERMAL CLIMATES
- HUMID CONTINENTAL, WARM SUMMER (Da)
- HUMID CONTINENTAL, COOL SUMMER (Db)
- SUBARCTIC (Dc, Dd)

POLAR CLIMATES
- TUNDRA (ET)
- ICE CAP (EF)

UNDIFFERENTIATED HIGHLANDS
- H
- EXTENSIVE UPLANDS SHADED

Scale at latitude 35°

Source: Trewartha, 1961

Latin America

Latin America contains several climatic zones from desert to tropical wet (see box 3.3). When asking about the region's climate problems, observers from outside Latin America are likely to focus on El Niño. Even if they do not know what El Niño really is, they probably know that coastal Peru is El Niño's "ground zero." Like North Americans, Latin Americans from the Rio Grande River to Tierra del Fuego constantly have to cope with a wide range of climate anomalies and climate-related impacts:

BOX 3.3
Climate Zones of Latin America

Latin America spans a vast range of latitudes and contains important high-elevation mountain ranges and, as a result, has a wide variety of climates. It is the only southern continent to reach high latitudes. Its broadest extent is in the equatorial zone; thus, tropical conditions prevail over the larger portion of the region. Mexico and Central America are affected by the penetration of cold fronts and tropical cyclones over the Atlantic and Pacific Oceans, whereas the Atlantic coast of South America is mostly free of high-intensity tropical storms.

Atmospheric circulation and cold ocean currents have remarkable influence on the weather and climate in the southern part of the region—giving origin to the Peruvian, Atacama, and Patagonian deserts, which receive less than 100 mm of mean annual precipitation. The cold Humboldt ocean current—which flows northward along the west coast of South America—brings to the coasts of Ecuador, Peru, and Chile large masses of phytoplankton that originate in the Antarctic Ocean, supporting one of the world's richest fisheries. This process is interrupted by an occasional appearance of warmer waters caused by the weakening of the westward flowing surface winds (e.g., El Niño) a phenomenon that brings heavy rains to an otherwise arid west coast of South America.

Although Latin America is characterized largely by humid, tropical conditions, important areas (e.g., northeastern Brazil) are subject to droughts and floods, and others are affected by freezes.

Source: www.grida.no (Global Resource Information Database Arendal)

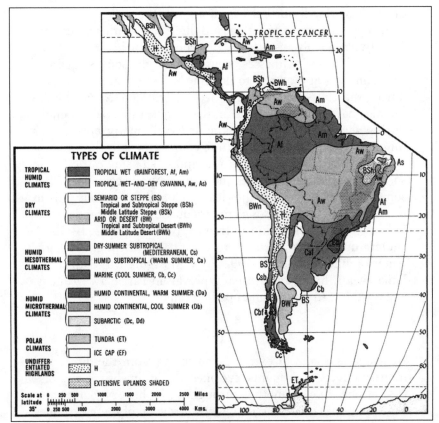

Source: Trewartha, 1961

droughts, floods, fires, hurricanes, coastal storms, and infectious disease outbreaks.

Central Americans, for example, face the risk of hurricanes each year from June to November. In the 1990s, there were several devastating—some might say blockbuster—hurricanes in the Gulf of Mexico, the Caribbean Sea, and the tropical Atlantic: Georges, Andrew, and Mitch. In one of the hemisphere's deadliest climate-related catastrophes, an estimated 20,000–50,000 Venezuelans lost their lives in December 1999 as a result of rapid-onset mudslides that resulted from heavy rainfall.

When an El Niño is under way, several parts of Latin America are adversely affected: parts of Central America are affected by drought or flood; northern Peru and southern Ecuador suffer from torrential rains and floods; Bolivia and southern Peru are usually, but not always, affected by drought; northeast Brazil (the famous *Nordeste*) witnesses drought while the southern part of the country is at high risk to flooding. These situations occurred during the 1997–1998 El Niño. Second-order effects of El Niño can also have adverse impacts on the environment. For example, during severe drought in northeast Brazil, many *nordestinos* migrate to major cities around the country or into the Amazon rainforest, where they resort to farming or raising cattle. To do so, they clear the land of trees and other natural vegetation. After some time, they abandon the cleared plots because poor soils in the rainforest cannot sustain crop production and torrential rainfall leaches the remaining nutrients from the exposed soils. A cholera outbreak in Peru led to a pandemic throughout Latin America during the 1991–1992 El Niño (figure 2.4). During La Niña events, heavy rains usually prevail in Paraguay, Uruguay, and northern Argentina and flooding can occur.

Australia

Australia is a land of extremes, suffering from the direct and indirect effects of droughts, cyclones, floods, bushfires, and disease at one time or another throughout its history. Sometimes these extremes occur in the same year. Several of these extremes have been correlated with either El Niño or La Niña episodes (e.g., Flannery, 1995; Nicholls, 1986). Australia could be considered the western Pacific's ground zero for El Niño's impacts and for the impacts of the related, seesaw-like, pressure patterns across the tropical Pacific known as the Southern Oscillation. From 1991 to 1995, Australia suffered from a five-year drought that some researchers blamed on El Niño. Agricultural production and ranching activities were adversely affected. A meteorological drought reappeared during the 1997–1998 event, but agricultural production was less adversely affected than earlier in the decade because of timely rains in the midst of drought. Severe drought with negative impacts on agriculture prevailed once again during the development of El Niño conditions in 2002–2003 (Meinke et al., 2003).

Australia is the globe's most arid inhabited continent, causing Australians to note that they are living in a sunburnt country (see box 3.4). It is plagued by a threat of bushfires. Many new plans have been put into effect since the 1983 Ash Wednesday

BOX 3.4
Climate Zones of Australasia

The region's climate is strongly influenced by its oceanic setting. Northern Australia, lying just south of the Western Pacific oceanic "warm pool," experiences tropical conditions, with a summer monsoon. Key climatic features include tropical cyclones and monsoons in northern Australia; migratory mid-latitude storm systems in the south, including New Zealand. An important feature of the region's climate is the ENSO phenomenon, which causes high year-to-year variability, especially of rainfall in northern and eastern Australia. The variability of Australian rainfall and runoff is among the greatest in the world: two to four times those of northwestern Europe and North America for the same climatic zones.

Tropical cyclones (averaging six per year) are a major concern for northern coastal regions. Cyclones can track southward as far as New Zealand, bringing very high rainfalls. Western and central Australia experience generally clear, dry conditions owing to large-scale downward motion of the atmosphere (e.g., subsidence).

Because of the size of Australia, the rain-bearing weather systems progressively dry out as they penetrate inland, resulting in a very arid central desert region. Australia is the driest populated continent; two-thirds of the land is classified as arid or semi-arid. Compared with other nations, the frequency and duration of drought are extreme, although these are interspersed with sequences of above-average rainfall. In contrast, New Zealand has a maritime climate; few places are more than 100 km from the coast. Rainfall is relatively well-distributed by region and season, and the influence of the ENSO phenomenon on year-to-year variations is less than in Australia.

Source: www.grida.no (Global Resource Information Database Arendal)

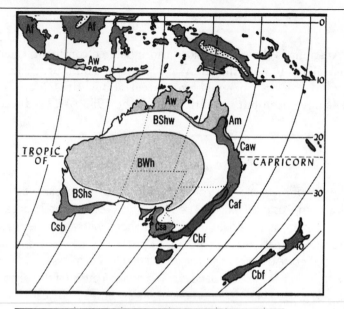

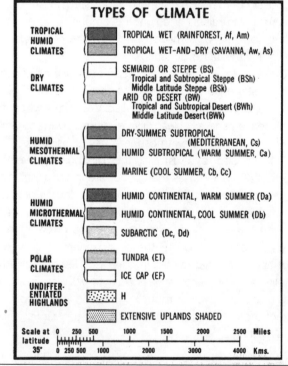

TYPES OF CLIMATE

TROPICAL HUMID CLIMATES
- TROPICAL WET (RAINFOREST, Af, Am)
- TROPICAL WET-AND-DRY (SAVANNA, Aw, As)

DRY CLIMATES
- SEMIARID OR STEPPE (BS)
  - Tropical and Subtropical Steppe (BSh)
  - Middle Latitude Steppe (BSk)
- ARID OR DESERT (BW)
  - Tropical and Subtropical Desert (BWh)
  - Middle Latitude Desert (BWk)

HUMID MESOTHERMAL CLIMATES
- DRY-SUMMER SUBTROPICAL (MEDITERRANEAN, Cs)
- HUMID SUBTROPICAL (WARM SUMMER, Ca)
- MARINE (COOL SUMMER, Cb, Cc)

HUMID MICROTHERMAL CLIMATES
- HUMID CONTINENTAL, WARM SUMMER (Da)
- HUMID CONTINENTAL, COOL SUMMER (Db)
- SUBARCTIC (Dc, Dd)

POLAR CLIMATES
- TUNDRA (ET)
- ICE CAP (EF)

UNDIFFERENTIATED HIGHLANDS
- H
- EXTENSIVE UPLANDS SHADED

Scale at latitude 35°
0 250 500 1000 1500 2000 2500 Miles
0 250 500 1000 2000 3000 4000 Kms.

Source: Trewartha, 1961

fires threatened towns near Melbourne. However, in late December 2001 and early January 2002, bushfires threatened the outskirts of Sydney. These fires could not be blamed on El Niño but instead were linked to lightning and arsonists. Although the country had developed sophisticated plans for reducing fire risk, the most recent fires were so large and widespread that they overtook the ability of the firefighters to cope effectively. Although no lives were lost in the 2001–2002 fires, large numbers of animals perished and hundreds of buildings and large expanses of vegetation were destroyed by flames.

Asia

To treat Asia as a unit would be extremely misleading to those unfamiliar with the continent. Asia encompasses a broad expanse of territory that is divisible into smaller geographically based regions with diverse topographical and climatic features: Western, Central, South, Southwest, East, and Southeast Asia (see box 3.5). Most people are aware of the monsoons in Asia and the important life-sustaining rains that they bring to countries from Pakistan to China. Problems arise, however, when either the monsoon is too weak or too intense. Both extremes adversely affect food production, and the floods can also destroy vulnerable infrastructure, villages, and rice paddies and worsen existing health conditions. Until the mid-twentieth century, hunger was chronic among large portions of the population and famines were not uncommon. For the Indian subcontinent, droughts seemed to accompany El Niño events (Davis, 2001). In the past two decades, however, that apparent relationship has not materialized, much to the benefit of Indian farmers.

Recurrent river flooding, droughts, and climate-related infectious disease outbreaks plague Bangladesh. It has also been in the path of devastating cyclones. A 1970 cyclone caused more than 300,000 deaths in that country, in what was then East Pakistan. Many of the country's drought and flood extremes have been correlated to El Niño and La Niña events, providing a level of predictability of these extremes and hope for climate-related disaster reduction.

Pakistan depends in large measure on water for irrigation to

BOX 3.5
Tropical and Subtropical Asia

The climate of tropical and subtropical Asia (A and C) is dominated by the two monsoons: the summer southwest monsoon (May to September), and the winter northeast monsoon (November to February). The monsoons bring most of the region's precipitation and are the most critical climatic factor.

As a result of the seasonal shifts in weather, a large part of tropical and subtropical Asia is exposed to annual floods and droughts. The average annual flood covers vast areas throughout the region; in India and Bangladesh alone, floods cover 7.7 million ha and 3.1 million ha, respectively. At least four types of floods are common: riverine flood, flash flood, glacial lake outburst flood, and breached landslide-dam flood. Flash floods are common in the foothills, mountain borderlands, and steep coastal river basis.

Tropical and subtropical cyclones are also an important feature of the weather and climate in parts of tropical Asia. Two core areas of cyclogenesis (where cyclones are born) exist in the region: one in the northwestern Pacific Ocean, which particularly affects the Philippines and Vietnam, and the other in the northern Indian Ocean, particularly affecting Bangladesh.

In the megacities and large urban areas, high temperatures and heat waves also occur. These phenomena are exacerbated by the urban heat-island effect and air pollution.

El Niño has an especially important influence on the weather and inter-annual climate and sea level variability, especially in the western Pacific and northern Indian oceans. The influence of Indian Ocean sea surface temperature on the large-scale Asian summer monsoon and hydrological cycle and the relationship between Eurasian snow cover and the Asian summer monsoon are also important factors.

Temperate Asia. Climate differs widely within temperate Asia (B and D). The region has a tropical monsoon climate in the far south; a humid, cool, temperate climate in the north; and a desert climate or steppe climate in the west and northwest. In the rest of the area—where most of the population of the region is concentrated—a humid, temperate climate prevails.

Temperate Asia is composed of three regions: so-called monsoon Asia, excluding its tropical subregion; the inner arid/semiarid regions (B), and Siberia (West Siberia, East Siberia, and Far East [D]), covered largely by boreal forests and tundra. Tropical cyclones (typhoons) frequent the coastal regions. Inner Siberia, with a mean monthly temperature in January of about

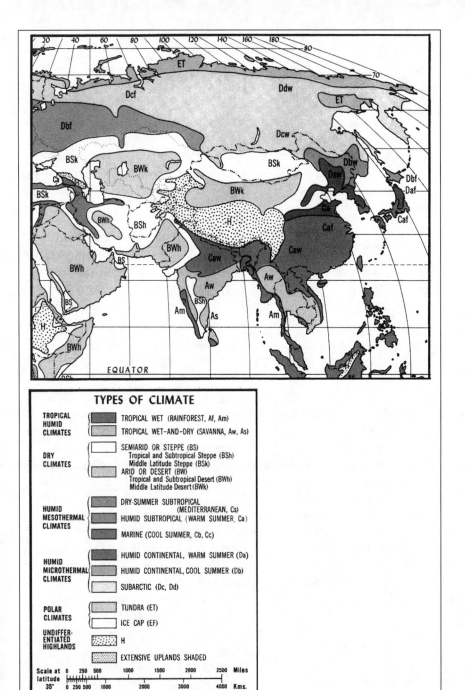

TYPES OF CLIMATE

TROPICAL HUMID CLIMATES		TROPICAL WET (RAINFOREST, Af, Am)
		TROPICAL WET-AND-DRY (SAVANNA, Aw, As)

DRY CLIMATES		SEMIARID OR STEPPE (BS)
		  Tropical and Subtropical Steppe (BSh)
		  Middle Latitude Steppe (BSk)

		ARID OR DESERT (BW)
		  Tropical and Subtropical Desert (BWh)
		  Middle Latitude Desert (BWk)

HUMID MESOTHERMAL CLIMATES		DRY-SUMMER SUBTROPICAL (MEDITERRANEAN, Cs)
		HUMID SUBTROPICAL (WARM SUMMER, Ca)
		MARINE (COOL SUMMER, Cb, Cc)
HUMID MICROTHERMAL CLIMATES		HUMID CONTINENTAL, WARM SUMMER (Da)
		HUMID CONTINENTAL, COOL SUMMER (Db)
		SUBARCTIC (Dc, Dd)
POLAR CLIMATES		TUNDRA (ET)
		ICE CAP (EF)
UNDIFFER-ENTIATED HIGHLANDS		H
		EXTENSIVE UPLANDS SHADED

Scale at latitude 35°: 0 250 500 1000 1500 2000 2500 Miles; 0 250 500 1000 2000 3000 4000 Kms.

Source: Trewartha, 1961

-50 °C (-58 °F), is the coldest part of the Northern Hemisphere in winter; this area is called the "cold pole." On the other hand, extremely dry, hot climate prevails in the Taklamakan Desert.

The Tibetan plateau (H), rising to the troposphere, strongly influences the atmospheric general circulation over the region, both thermally and dynamically. Development of polar frontal zones and cyclones in temperate Asia is closely connected to activities associated with the westerly jet stream and the East Asian monsoon, both of which are significantly affected by the plateau.

Middle Eastern and Arid Asia (B). Two-thirds of the region can be classified as hot or cold desert. In the northern part of the region, a steppe climate prevails, with cold winters and hot summers. A narrow zone contiguous to the Mediterranean Sea is classified as a Mediterranean zone, with wet and moderately warm winters and dry summers. Permafrost zones exist in high mountain areas in the southeast part of the region.

Source: www.grida.no (Global Resource Information Database Arendal)

produce many of its crops for domestic use and for export. It has served as a refuge for Afghans who fled from the multiyear drought in the early 2000s or the Taliban regime, or both. This created major problems for the Pakistani government, because it had to provide for the needs of the large number of displaced Afghans on its soil. As a result of the drought in Central and Southwest Asia (Afghanistan, Iran, and Pakistan), food shortages and even famine occurred in some locations, prompting an urgent need for additional humanitarian food assistance.

Indonesia, the Philippines, and the countries in Indochina (Vietnam, Laos, and Cambodia) are affected by El Niño- and La Niña-induced droughts and floods, with these extremes occurring in different parts of these countries at the same time. In addition, parts of Indonesia and Malaysia suffer from widespread forest fires during El Niño, with resultant haze throughout the region.

Many publications have been written over the centuries about the impacts of China's droughts and floods (Chu, 1954; Wang, 1981). Despite great strides in China toward economic development, many parts of the country remain vulnerable to high-impact weather and climate extremes. For example, the Yangtze River basin floods in the summer of 1998 displaced an estimated one-fourth of the country's billion-plus population. However, unlike during floods earlier in the twentieth century, the government had the resources (i.e., the army) to move most of the potential victims out of harm's way. This particular flood was so devastating and costly that China reviewed its land-use policies in the river basin, as well as its weather forecasting practices and regional climate services capabilities (Ye and Glantz, 2002). As bad as it was, however, it was not the worst climate-related disaster in twentieth-century China. Other more deadly floods and drought-producing famines occurred as recently as the 1960s.

Then there's the "roof of the earth," the Chinese and Tibetan mountains, where most people see snow, ice, glaciers, mountains, and a perennially harsh climate. Yet, such is not the case. Geoffrey Smith (1987) quoted Kingdon Ward, who, while on a search there for varieties of alpine plants, made the following firsthand observation in the early decades of the 20th century: "Though the summers are wetter, the autumn sunnier, and the winters colder, the seasons are as distinct as they are in Great Britain."

Antarctica and the Arctic

In the past few decades, scientific research efforts have intensified for the Antarctic environment. The research was partly supported by various governments to maintain a political presence on the continent. However, concern about global warming and chemical emissions into the stratosphere have made the polar regions much more important for scientific research. One of the major greenhouse gases, the chlorofluorocarbons, has been implicated since the early 1970s in the thinning of stratospheric ozone. The surprising discovery of the Antarctic ozone hole in the mid-1980s helped to catalyze governments to pursue international negotiations to bring to a halt the use of ozone-eating chemicals (Litfin, 1994).

As previously noted, there is a looming concern about the possible disintegration of the West Antarctic Ice Sheet (WAIS).

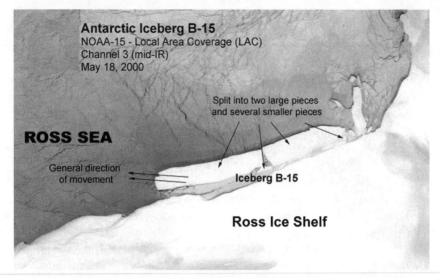

Antarctic Iceberg B-15
NOAA-15 - Local Area Coverage (LAC)
Channel 3 (mid-IR)
May 18, 2000

Split into two large pieces
and several smaller pieces

ROSS SEA

General direction
of movement

Iceberg B-15

Ross Ice Shelf

Figure 3.2. Antarctic iceberg B-15 broke away from the Ross ice shelf in mid-March 2000. It was among the largest pieces ever observed, at an estimated 11,000 square kilometers, or the size of Connecticut. (Source: NOAA)

In fact, signs of disintegration of the Antarctic ice mass have begun to appear, as large segments approximating the size of the U.S. states of Delaware and Connecticut have already broken away. Figure 3.2 shows an iceberg on the Ross ice shelf in Antarctica breaking up into several pieces in mid-May 2000. Several other state-size icebergs have also broken away from the Antarctic ice shelf.

The Vostok ice core taken at the Soviet (now Russian) research station in East Antarctica has provided climate researchers with a long-term historical record of global climate (and atmospheric gases) that spans about 420,000 years before the present.

Scientists expect that for each degree of warming in the mid-latitudes, a $3–4°C$ (about $5–7°F$) warming can be expected in the polar regions. It appears that signs of global warming are already showing up in the Arctic, as manifested by the thinning of sea ice in the polar region. This thinning has implications for the indigenous populations in the circum-Arctic region, such as the Inuit in northern Canada. In addition, there is an international political dimension to the thinning ice: as the sea ice recedes, so does the territorial extent of Canadian sovereignty. This means that ice territory presently controlled by Canada will revert to the international community and to the rule of the international Law of the Sea.

Africa

Climate variability from season to season and year to year has been a major problem for regions, countries, and inhabitants in each of the distinct agro-climatic subregions of Africa (see box 3.6). However, the West African Sahel droughts drew worldwide attention to the extent of vulnerability to climate of ecosystems and inhabitants on the continent. Major droughts in the Sahel occurred when the region was under colonial rule in the 1910s and again in the early 1940s. Its most recent devastating drought occurred from 1968 to 1973 and, depending on the measures used, some say a period of increased aridity extended to 1985. Drought conditions continue to adversely affect the economic development prospects of African countries, both above and below the Sahara. For example, recurrent drought episodes and desertification affect Nigeria (Watts, 1983; Olori, 2002), the continent's most populous country.

Two countries in southern Africa, South Africa and Zimbabwe, are able to produce food in surplus, allowing them to export maize, for example, throughout the region. They have favorable soils and climate conditions much of the time. However, they do suffer from severe drought conditions every so often. Many, not all, of these droughts have been statistically linked to the onset of El Niño events. South African governments have been concerned throughout the twentieth century about the negative impacts of drought on both agricultural and rangeland production (e.g., South African Drought Commission, 1922; Monnik, 2001). Although drought is a factor in food production problems, politics also play a key role. For example, today Zimbabwe is at a high risk of food shortages and possible famine, given recent climate anomalies of floods followed by drought, in addition to its unstable political situation and ad hoc land planning activities. If the predicted El Niño occurs in late 2002–2003 (Schmid, 2002), a bad situation will only be made worse. Mozambique usually suffers from drought during El Niño episodes. The country saw

---

BOX 3.6
Climate Zones of Africa

The African continent is characterized by several climatic regimes. All parts of the continent, except the Republic of South Africa, Lesotho, and the Mediterranean countries north of the Sahara, have tropical climates. The tropical climates can be divided into three distinct zones: wet tropical climates, dry tropical climates, and alternating wet and dry climates.

Summer rainfall maxima, which are dominant over most of Africa, are controlled primarily by the Inter-Tropical Convergence Zone (ITCZ). Over land, the ITCZ tends to follow the seasonal march of the sun and oscillates between the fringes of the Sahara in northern hemisphere summer and the northern Kalahari desert in the southern summer. The latitude zones of these arid and semi-arid deserts demarcate the tropics from the subtropics. Rainfall in the subtropics is modulated by mid-latitude storms, which may be displaced Equator-ward in winter. Further alterations of these broad patterns is provided by natural features such as lakes and mountains, and by the influence of ocean currents. The poleward climatic extremes of the continent have extra-tropical influences.

Rainfall over Africa exhibits high variability in space and time. Mean annual rainfall ranges from as low as 10 mm in the innermost core of the Sahara to more than 2000 mm in parts of the equatorial region and other parts of west Africa. The rainfall gradient is largest along the southern margins of the Sahara—the region known as the Sahel—where mean annual rainfall varies by more than 1000 mm over about 750 km. Drought is a normal feature of Africa's climate and it is acknowledged that its recurrence is inevitable.

Source: www.grida.no (Global Resource Information Database Arendal)

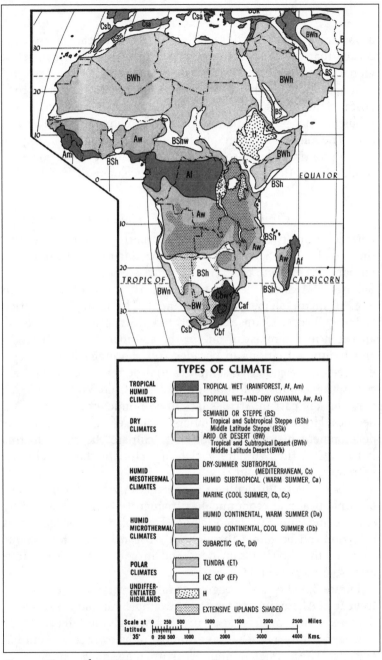

TYPES OF CLIMATE

TROPICAL HUMID CLIMATES
- TROPICAL WET (RAINFOREST, Af, Am)
- TROPICAL WET-AND-DRY (SAVANNA, Aw, As)

DRY CLIMATES
- SEMIARID OR STEPPE (BS)
  - Tropical and Subtropical Steppe (BSh)
  - Middle Latitude Steppe (BSk)
- ARID OR DESERT (BW)
  - Tropical and Subtropical Desert (BWh)
  - Middle Latitude Desert (BWk)

HUMID MESOTHERMAL CLIMATES
- DRY-SUMMER SUBTROPICAL (MEDITERRANEAN, Cs)
- HUMID SUBTROPICAL (WARM SUMMER, Ca)
- MARINE (COOL SUMMER, Cb, Cc)

HUMID MICROTHERMAL CLIMATES
- HUMID CONTINENTAL, WARM SUMMER (Da)
- HUMID CONTINENTAL, COOL SUMMER (Db)
- SUBARCTIC (Dc, Dd)

POLAR CLIMATES
- TUNDRA (ET)
- ICE CAP (EF)

UNDIFFERENTIATED HIGHLANDS
- H
- EXTENSIVE UPLANDS SHADED

Scale at latitude 35°: 0 250 500 1000 1500 2000 2500 Miles / 0 250 500 1000 2000 3000 4000 Kms.

Source: Trewartha, 1961

devastating floods in 2000 and 2001. African leaders are concerned that the climate situation will worsen with the onset of global warming.

It appears that sub-Saharan Africa has become the proverbial poster child for food crises and especially food-related nutritional problems that put populations at risk for climate-related anomalies. The following paragraphs focus on the challenges facing the African continent but also offer a potent example of the powerful conditioning effect that climate has on development.

Climate and African Development

For the entire twentieth century and perhaps more so toward the end of it, sub-Saharan Africa has been in a "C" of trouble. In January 1984, *Time* magazine devoted an entire issue to sub-Saharan Africa. The cover portrayed the head of an African woman in the shape of the continent. The issue was entitled "Africa's Woes: Coups, Conflict and Corruption." This issue appeared in the midst of one of the continent's worst drought and famine situations in decades. Yet the magazine's articles devoted only a couple of sentences to drought! Clearly, conflict, corruption, coups, colonialism, and the Cold War have each played a key part in Africa's lack of economic progress over the decades. However, a better understanding of Africa's economic development history demands that "climate" be added to this list of "C's" (i.e., climate variability, climate change, climate extremes, and seasonality).

It is not very difficult to argue that African development, both political and economic, has frequently been influenced by climate and climate-related factors. Examples in support of this assertion can be found throughout African history for various parts of the African continent and in contemporary African affairs (Anderson and Grove, 1987).

Figure 3.3 shows the degree of rainfall variability from year to year in Africa. In arid lands, for example, rainfall is skewed to dryness, which means that relatively few high rainfall events are averaged out by a much larger number of below-average rainfall episodes. The average rainfall departure map shows a horseshoe-shaped region in sub-Saharan Africa where rainfall is highly

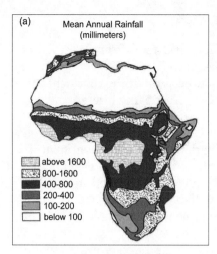

(a) Mean Annual Rainfall (millimeters)

- above 1600
- 800-1600
- 400-800
- 200-400
- 100-200
- below 100

Figure 3.3.

a) Climatological mean annual rainfall distribution over Africa.

b) The distribution of the average departure of rainfall in Africa from the long-term mean expressed as a percentage.

c) Map depicting African countries facing extreme food shortages in 1984.

(Source: WMO, 1984)

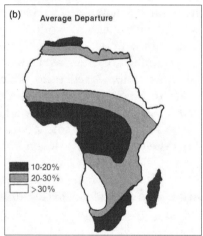

(b) Average Departure

- 10-20%
- 20-30%
- >30%

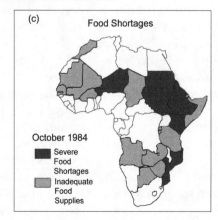

(c) Food Shortages

October 1984

- Severe Food Shortages
- Inadequate Food Supplies

variable from one year to the next. It extends from West Africa across the continent to Ethiopia, down the east coast of Africa, and then westward across the southern tip of the continent into Namibia. Forecasts of rainfall at the seasonal to interannual, and even decadal, scales become extremely important in this region.

Furthermore, a rainfall map for Africa does not tell the whole story about what life is like in various parts of the continent. To people far away on another continent, rain means water. Water means agriculture. Agriculture means food. Food means life. However, the timing of the rainfall makes a big difference in the amount of precipitation that is effectively available for use by plants, animals, and people.

Africa is divided into several ecological regions, from the rainforests to the hyper-arid Saharan desert. Some regions have one rather short annual rainy season, while others have two robust rainfall periods within a calendar year. Some inhabitants can count on reliable rainfall for their agricultural activities. Others have turned to irrigation practices to produce food for subsistence, the market, or export. Still others cannot rely on the irregular rains to fall on specific plots of soil and have resorted to the keeping of livestock. Thus, to rely solely on average yearly rainfall amounts for development purposes in such regions is very misleading.

Climate is only part of the problem in the region, however. The ways governments and individuals decide to use land in a given region is another part of the problem. Many of Africa's land-use practices that are destructive in the long term can remain hidden during periods of good rains. For more than half a century, people have written about destructive land-use practices in various parts of sub-Saharan Africa: tree cutting to open up forested areas to farming or for the use of wood to construct dwellings and protective fences; overgrazing of grasslands by large herds of goats and cattle; widespread drilling of bore holes and deep wells in arid areas; firewood gathering; dung collection, and so forth. When a drought or extended dry spell occurs, inappropriate practices are exposed.

Many people in Africa have been enticed, if not forced, to move into areas that are increasingly marginal for agricultural production. Various schemes have been proposed to increase rainfall and water resources on the African continent. Yet, while researchers and decision makers have been busy devising ways to bring marginal lands into production, people have been misusing and abusing the productive land. Much of that misuse has been caused by forces beyond the direct control of local inhabitants, who are eking out a subsistence living from the soil. Either they have been encouraged by their governments to exploit new lands for agricultural or grazing uses or they have been responding to the direct and indirect pressures from natural population increases.

Making Africa's future prospects for economic development even more difficult than in the past, African leaders and researchers among others, became increasingly concerned in the early 1990s about the potential impacts of global warming on the continent. Predictions have been made about changes in precipitation, seasonality, and sea level rise. Hence, considerable attention has been paid to speculation about climate change impacts in Africa based on historical accounts, computer-modeling activities, and on worst-case scenarios. The good news is that many African researchers have become engaged in international and regional activities centered around climate and climate-related research (Ogola, 1997; Ominde and Juma, 1991; GLOBE, 2000).

Some Climate Concerns of Africa

Africa's climate concerns center on three items: early warning systems, food security, and global warming. African governments and nongovernmental organizations (NGOs) working in Africa are concerned about the lack of a universally accepted reliable early warning system for extreme climate events on the continent. Several governments, UN agencies, and NGOs have sought to develop climate-related early warning systems for drought, flood, disease, and famine. Often, even within the same country one may be able to identify several early warning systems at work, each identifying its own set of indicators to monitor the same climate-related problem.

Food security is a constant concern to African leaders, to NGOs, and to other humanitarian and donor organizations because chronic malnutrition and hunger plague many people throughout the continent, even under normal climate conditions. Hunger is not confined to one location or one ethnic group. All the agencies involved with food issues want to avoid a worsening of the existing food security situation—a worsening that could result from drought-related impacts, e.g., water shortages, crop failure, rangeland degradation, infectious disease outbreaks, environmental degradation, and food- and water-related conflicts.

Global warming will affect the location, frequency, and amount of precipitation and evaporation on the African continent as well as increase the likelihood of saltwater intrusion in coastal areas and aquifers as a result of sea level rise. Global warming concerns in Africa revolve around an ability to sustain food production, avoid potential environmental degradation, minimize out-migration, and reduce the risk of hunger and starvation. Water and food crises appear to have become expected occurrences in sub-Saharan Africa and have apparently been overshadowed by the HIV/AIDS situation, which has reached crisis status in Africa.

Real-Life Climate Issues in Africa

Every year there are likely to be several climate and climate-related problems confronting Africans. Some recent examples serve to suggest the difficulties that decision makers interested in the well-being of Africans have to face.

- The impacts of recurrent drought in the West African Sahel and Ethiopia generated global awareness of how the interplay between land-use practices and high levels of interannual climate variability can lead to desertification. Understanding the climate characteristics of a region can help to minimize such land degradation.
- During the 1997–1998 El Niño event, Kenya was flooded by heavy rains. Roads were washed out (including the main route between Nairobi and Mombasa), human and livestock diseases (e.g., Rift Valley fever) spread to epidemic levels, and international assistance was needed for medicines and food. The floods were blamed on the El Niño, but not every El Niño has been associated with flooding in Kenya. However, in this instance, forecasters in the

Northern Hemisphere failed to take into account the influence of anomalously warm waters in the western Indian Ocean off the coast of eastern Africa. Apparently, this influenced rainfall conditions in Kenya more so than did the El Niño in the distant central Pacific Ocean.

- Since the late 1960s, Africa's Lake Chad, in the midst of the Sahara Desert, has been drying up. It has dropped in area from about 20,000 sq km then to about 1,200 sq km today. The lake had been a source of fish and agricultural activities around its edges. Today, settlements that had been on the shore of this large but shallow lake are now many tens of kilometers from it. Levels of Lake Chad have fluctuated throughout history. There is no major human disturbance of the rivers that flow into it. Thus, one can view the desiccation of Lake Chad as a natural process that may have been accelerated by global warming. (In fact, many inland seas [with no outlet to an ocean] around the world have been drying out; some have already disappeared [e.g., Lopnor in western China]. Inland seas, however, appear to be affected as much, if not more, by river diversions as by natural fluctuations in the climate system.)

- The major source of the Nile River in Egypt is the Blue Nile, whose headwaters are in Ethiopia. The Ethiopians to date have not undertaken major diversions from the Blue Nile to meet their own national development needs. The river has been untouched because of the country's low level of economic development and because of recent decades of internal and international conflicts on Ethiopian soil. However, given that Ethiopia has been plagued by recurrent, drought-related famines in recent decades, and given its awareness of the likelihood of El Niño–related droughts and the specter of global climate change, the government has begun to consider water from the Blue Nile as a possible source to mitigate the impacts of future water shortages in the country. Out of deep concern that the Ethiopian government may choose to tap the Blue Nile at some time in the future, the Egyptian authorities have threatened Ethiopia with war if they attempt to divert waters from the Blue Nile (Weinthal, 2001).

- Central Africa's dry and wet forests are being cut down for tropical timber exports to developed countries. The companies involved in deforestation are not African; some come from as far away as Malaysia. These companies feel no need to protect their own country's environment, let alone the African environment. This may not be just a local environmental problem for some parts of central Africa. Researchers have recently argued that deforesting central Africa could have a major negative influence on rainfall in southern Africa, a region which is already subjected to high rainfall variability and wet and dry extremes (Forest Monitor, 2001).
- There is an ongoing competition between using agricultural areas to grow subsistence crops for domestic consumption or cash crops for export. This competition can be found in many countries: in the Gezira (cotton) scheme of the Sudan and in the export of *qat* from the Hararghe region in Ethiopia to Yemen, for example. Although in a perfect world this might make sense—sell cash crops and import food with the earnings—we do not have a perfect world. As another example, a few West African governments were discovered to have been growing and exporting cash crops in the early 1970s during the five years of drought, food shortages, and famines in the region (e.g., Lofchie, 1975).

This brief glimpse of how climate generates problems for Africans—peasants as well as presidents—shows that climate-society-environment interactions are truly dynamic and in many instances, health if not life threatening. The looming prospect of global warming is just one more climate-related burden placed on African societies that are already at risk from existing climate-related problems.

FOUR

WHAT IS CLIMATE AFFAIRS?

The concept of climate encompasses a wide range of average meteorological conditions, as well as extremes, fluctuations, variability, and changes in temperature, precipitation, and pressure. "Climate affairs" looks at climate within its human context. It describes a multifaceted approach to understanding and managing the many ways that climate, in its broadest sense, influences human activities and environmental processes. It has emerged in recent years as a result of the growing realization that humans affect climate as much as they are affected by it.

Unlike, say, "climate science" or "climate policy," the term "climate affairs" encompasses a wide range of issues and disciplines related to atmospheric properties and processes that affect climate. It merges science, policy, and ethics in the study and management of such phenomena as chemical emissions into the atmosphere, land-use processes, energy supply and demand, water resource availability, food production and availability, public health conditions, public safety as affected by climate-related hazards, and so on. It draws upon insights from several academic disciplines in the physical, biological, and social sciences, and in the humanities. It also involves socioeconomic sectors of society, such as transportation, tourism, clothing manufacturing, commerce, and trade.

The field of climate affairs was developed in a conscious attempt to put climate and climate-related factors on the list of items that decision makers normally take into consideration. It is also important to society's educators. If those in a position to formally or informally teach others can be reminded of the various direct and indirect ways that atmospheric processes can influence the issues in which they have an interest, they are more likely to refer explicitly to climate and climate-related issues as an integral part of their education or training activities.

Climate affairs does *not* attempt to trace all of society's ills to climate-related factors. There are times when the consideration of climate proves to be very important to decision makers, while at other times it may be of little relevance. In any event, climate needs to be taken into account.

Many decision makers truly do not understand many of the obvious, let alone subtle, ways that climate anomalies can affect the activities for which they are responsible. They are simply unaware of potential influences of climate on society. The goal is to make them aware of such influences so that in the face of future anomalies they have the option to pursue proactive strategies and not just rely on reactive ones.

To meet the societal need for a better understanding of the many ways that climate variability and change (on a variety of time scales from seasons to centuries) affect ecosystems and the affairs of people and nations, we need a multifaceted perspective. Armed with such a perspective, decision makers in all socioeconomic sectors of society will be better informed, as will the general population.

Climate affairs includes the following component fields:

- *Climate science* is a general description of the physical climate system's components and the inclusion of human activities as a forcing factor of local to global climate.
- *Climate impacts* refers to the impacts of atmospheric processes on both societies and ecosystems. It also includes the impacts on the atmosphere of human activities and ecological processes.
- *Climate politics* (domestic and international) is the process

required to produce climate-related regulations and laws.
- *Climate policy and law* encompasses the legal and regulatory aspects of climate-society-environment interactions.
- *Climate economics* relates to the benefit and cost assessments of climate anomalies and society's attempts to prevent, mitigate, or adapt to their impacts.
- *Climate ethics and equity*, the ethical aspects of climate-related issues, have often been overlooked in past assessments of climate-society-environment interactions.

Each of these aspects of the notion of climate affairs is briefly described in the sections that follow.

Climate Science

Climate science provides the foundation for a reliable understanding of the climate system and its interactions with and impacts on societies and the environment. An introduction to the workings of the global climate system can improve how we deal with present-day and future climate conditions. Although it is difficult to identify here all aspects of the climate system, a glimpse of key aspects will prompt people to seek more information at their desired level of scientific detail.

From a traditional climatological perspective, the climate system encompasses the geophysical and atmospheric aspects, characteristics, and processes that combine to produce the climate. These aspects include but are not limited to

- the oceans,
- vegetation,
- topographical features,
- volcanoes,
- clouds,
- the sun,
- ice and snow cover,

and so forth. The science of climate as a course of study includes various atmospheric processes involved in global, regional, and local climate variability, fluctuations, change, extremes, and seasonality. It also includes the study of the atmospheric and ecological processes that lead to long-term environmental change, as happens when a region shifts from forest to savanna or from savanna to desert. Understanding the atmosphere—its constituents and its processes—and the various ways that it is or can be influenced by oceanic and biological changes in the marine and terrestrial environments has been a primary focus of traditional climate science research and of the climate forecasting community.

The objectives of climate science are at least threefold: to describe the structure and function of the global climate system; to improve the understanding of its components; and to show how human activities influence an otherwise physically and biologically driven system.

Figure 4.1 shows the components of the climate system. To those who are not physical scientists, it is a realistic, simplified view of climate and the factors that influence it. The interactions of these factors determine climates at all space and time scales.

However, figure 4.1 contains only the physical and biological components of the system. It does not include societal influences on the atmosphere. Landsberg (1956) wrote that "early students of climate were quite conscious of the fact that man's activities were likely to cause changes of climate. Thomas Jefferson . . . wrote in a letter dated at 'Monticello,' July 16, 1824, that climate surveys 'should be repeated once or twice in a century to show the effect of clearing [of land] and culture [settlements] towards changes of climate.'"

Of course, almost everyone now believes that human activities can alter, if they are not already altering, the workings of the climate system at the local scale, if not globally. Human activities are being accepted by a growing number of physical scientists as an integral part of the global climate system, because such activities are capable of forcing changes in otherwise naturally occurring atmospheric processes, from local to global scales. They do so by altering the natural balances among chemical constituents in the atmosphere and by land-use changes.

Human activities capable of influencing climate processes include the emissions of greenhouse gases and ozone-depleting

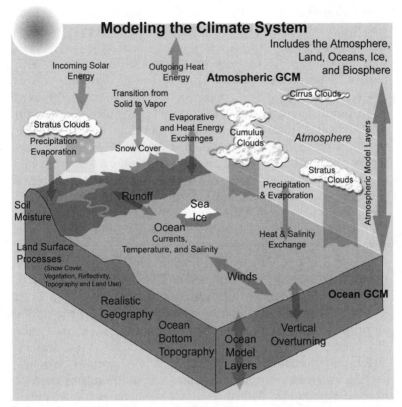

Figure 4.1. GCM refers to general circulation models. (After: National Assessment Synthesis Team, 2000, p. 15)

substances and land-use practices such as land clearing for cultivation, industrial processes, urbanization, livestock grazing of rangelands, tropical deforestation, and so on. Human activities affect urban climate through the urban heat island effect. This results from fossil fuel burning, heating and air-conditioning, and the large surface area covered by asphalt and concrete, which tends to retain and reemit heat slowly to the air. Human activities can also alter regional climatic conditions. From the large industrial plants in the Northern Hemisphere to suburban commuters to a peasant family in the tropics burning firewood to cook an evening meal, people everywhere are affecting atmospheric chemistry and, as a result, are affecting climate conditions at various space and time scales.

A good example of how human activities interact with natural processes to change climate at the regional scale is desertification. Desertification processes captured government attention in the early 1970s as a result of a multiyear drought in the West African Sahel. A debate developed as to whether the creation of desertlike landscapes in the region (and elsewhere on the globe) was due to a natural increase in aridity or to inappropriate land-use practices, including overgrazing of the landscape by excessively large herds, cultivating land where rainfall is often deficient for sustainable or reliable crop production, stripping land of its natural vegetative cover in order to farm, and collecting firewood and wood for construction purposes. Many researchers believed that human activities were the primary cause of the increased aridity (e.g., Charney, 1975). However, a reassessment of changes in the vegetative cover in this particular region using satellite images from the 1980s identified a natural ebb and flow of the Sahara Desert's edge (Tucker et al., 1991). The study suggested that there was an alternation between wet and dry periods over only a few decades, with associated changes in the amount and type of natural vegetation. These results do not mean that human activities do not have adverse effects on vegetation in the Sahel. They do mean that sorting out the interactions among human activities, climate fluctuations, and environmental changes requires considerable scrutiny before one can make a definitive attribution of cause.

Climate science now includes research from outside the physical sciences to take into account the fact that societies influence atmospheric processes and climate. Of course, major international efforts have now produced a large body of evidence indicating that carbon dioxide and other trace gases are warming the earth's atmosphere.

Figure 4.2 correctly highlights human activities as a key aspect of the climate system of the twentieth century. It gives human activities an elevated status and recognizes their proper role in the climate system as a forcing factor.

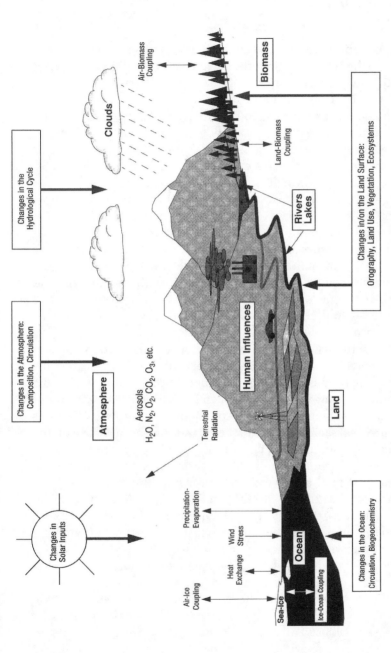

Figure 4.2. Schematic view of the components of the global climate system (bold), their processes and interactions (thin arrows) and some aspects that may change with climate change (bold arrows). (Source: Trenberth et al, 1996, p. 55)

Climate Technology

Human influences on climate have not always been inadvertent. Throughout history, societies have tried to devise ways to cope with what one writer has referred to as "problem" climates (Trewartha, 1961). In arid areas, for example, people have resorted to irrigation to bring water to potentially fertile but parched soils. Often the sources of water are not exactly where people choose to settle. Many ancient civilizations arose in arid environments where rivers and streams existed, seasonally if not perennially, as a result of snowmelt in high mountainous areas. Settlements originally appeared near those rivers and streams to compensate for the difficult climate conditions in the arid regions. Trees were planted in an attempt to modify local climate conditions.

In recent times, the invention of air-conditioning was viewed as a major breakthrough because pockets of temperate climate could be brought to inhospitable climates. Air-conditioning was expected to make people in areas with hot and humid climates relatively more energetic and as productive as those in the northern latitudes (Markham, 1947).

Over the past century, several climate modification schemes have been proposed for different parts of the globe, each with a different purpose: damming the Bering Sea; filling dry inland drainage basins in northeastern and southern Africa with water from the sea or rivers, respectively; planting rows of trees as wind breaks along the northern and southern edges of the Sahara, in the American Great Plains, and in China; cloud seeding monsoonal fronts to shift rainfall from wetter to drier inland areas of Brazil and West Africa; towing Antarctic icebergs to coastal arid countries such as Saudi Arabia or Argentina; laying asphalt on part of Libya's coastal desert to stimulate orographic (mountain) rain-producing effects (i.e., thermal mountains) over otherwise barren, unusable desert areas.

Figures 4.3 and 4.4 identify proposed climate modification schemes around the globe and in Africa, respectively. A number of the more spectacular schemes have been proposed for Africa. These have been seriously discussed at various times throughout the 1900s. For the most part, they were attempts to increase precipitation, streamflow, or water resources in arid and semiarid areas. A brief description of each and its intended purpose follows:

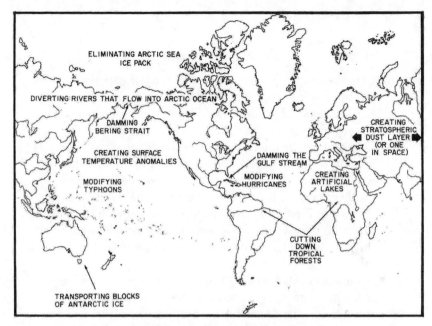

Figure 4.3. Engineering schemes that could be or have been proposed to modify or control the climate. (Source: Glantz, 1977, p. 311)

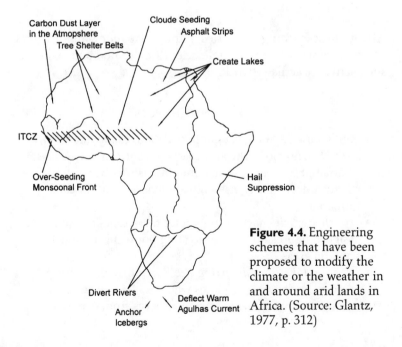

Figure 4.4. Engineering schemes that have been proposed to modify the climate or the weather in and around arid lands in Africa. (Source: Glantz, 1977, p. 312)

Temporary Changes

- *Cloud seeding* refers to putting silver iodide particles directly into the atmosphere, where they can serve as nuclei around which droplets of water can coalesce to form larger raindrops.
- *Overseeding the monsoonal front* was proposed as an attempt to shift a portion of the rainfall from normally wet regions to the arid interior of the continent at the southern edge of the Sahara. This would be accomplished (at least in theory) by oversupplying the air with nuclei so that raindrops would not become heavy enough to fall to the ground as rain until the front had moved farther inland.
- A variant of cloud seeding was the idea to put *carbon dust in the air* to create a heat source that would make the atmosphere unstable and more conducive to storm cloud formation.
- *Hail suppression* was designed to reduce the size and hardness of the frozen water that was destroying crops on tea plantations.

All of these seeding activities require frequent applications. There is, however, considerable scientific debate about whether such activities actually work.

Permanent Changes

- ESSO (now EXXON) sought in the 1960s to *lay alternating asphalt strips* along the arid coasts of continents. This was an attempt to create thermal mountains that would force moisture-laden air to rise high enough to result in cloud formation.
- *Tree shelterbelts* have been popular for decades and are designed to break up the desiccating hot winds in arid and semiarid areas.
- *Creating lakes and diverting the flow of rivers* was supposed to bring water to places where societies wanted to encourage development. Towing icebergs from Antarctica

was suggested as early as the 1880s to bring freshwater to arid areas.

- A suggestion was made to *deflect the direction of ocean currents* away from the wet, southeast coast of southern Africa to the arid southwest coast.

Although many of these schemes may work in theory, putting them into practice proves to be much more difficult. Nevertheless, from the perspective of the beginning of the twenty-first century, some of the proposals for climate modification appear to border on science fiction.

What these schemes highlight is that the climate system is in constant flux, as are most of its components. In the past, our ability to understand all of the simultaneously occurring interactions was limited by the reliance on "human computers" (i.e., statisticians) to carry out various calculations to identify correlations between causes and effects. That changed with the advent of satellites and computers. Satellite technology provides us with the opportunity to observe and monitor, over long periods of time, environmental and demographic changes taking place across large areas. Monitoring such changes on a global scale becomes more cost-effective than having to do so on the ground at many locations on the earth's surface. Satellites can provide information about potentially unwanted changes to the landscape, for example. Such an early warning can then stimulate the activation of "ground-truthing" measures in very specific locations to validate changes identified in satellite images.

Today, computers make billions of calculations in seconds, and our ability to run computer models of atmospheric interactions and the components that influence the earth's climates has sharply increased. Although such general circulation models have limitations, they do assist researchers in attempts to better understand the structures, functions, and interactions of the components within the climate system.

Just as physical scientists have slowly but increasingly come to recognize that human activities have become a notable forcing factor in atmospheric processes and climate, it is time for social scientists to recognize the importance of understanding the physical settings under which society must operate.

Climate Impacts

"Climate impacts" refers to the impacts of climate on both ecosystems and societies. Ecosystems include those that are managed (i.e., agricultural and rangelands) as well as those that are unmanaged (i.e., some forests, arid lands, marine, and wetlands). Ecosystems can be further subdivided into terrestrial and marine ecosystems. Societal impacts can be subdivided into the direct (first-order) and indirect (second- or third-order) impacts of climate variability, climate change, and extreme events on human activities and especially on the resources on which those activities depend.

Aside from the financial costs associated with a disaster, there is also a "misery factor" that must be taken into account. Misery refers to displaced populations; reduced nutritional intake; lost livestock, personal possessions, and pets; stress-related health impacts; and so on. An example of the climate-related misery in a developing country is shown figure 4.5. Ethiopians of all ages have to endure sitting in the hot sun for hours several times a month to receive a sack of donated grain to feed their families during a drought.

Various organizations have developed misery indexes to quantify the degree of misery at a given point in time. There is an annually issued *Forbes Global Misery Index* on the state of the global economy, an *LA Misery Index* for the city of Los Angeles, a country-specific financial misery index, a corporate misery index, and, more relevant to developing countries, a *UN Development Programme Misery Index*. The latter tries to incorporate qualitative aspects of life as well as quantitative ones. To the author's knowledge, none encompasses the misery aspect of climate-related societal impacts as seen, for example, in figure 4.5.

The study of climate impacts also includes the influences that societies can have on climate (e.g., land clearing, deforestation, urbanization, and greenhouse gas (GHG) emissions, especially carbon dioxide).

When it comes to climate impacts, the tendency for researchers as well as policy makers is to focus on hazards and disasters. However, it is important for societies to also assess

Figure 4.5. People patiently wait all day long in the hot sun for their turn to receive grain rations in the Hararghe region of Ethiopia during a serious drought in 1987. (Photo by M. H. Glantz)

climate as a resource. The climate resource includes not only favorable climate characteristics in a given region, but also the detailed information about climate conditions in the region of interest. A more effective knowledge of existing climate information (averages, extremes, modes, etc.) can reap tremendous benefits for those who choose to use it.

Impacts on Ecosystems

Climate affects marine and terrestrial environments. For example, changes in climate can affect the abundance and location of fish populations, thereby increasing or decreasing their availability to fishermen and natural predators. It is easy to show how climate variability, change, and extremes can affect the environment and therefore the health of particular fish populations, for good or ill. Fishing pressure caused by natural predators as well as by humans makes a fish population's survival even more difficult.

With regard to climate impacts on ecosystems, primary

concern centers on managed ecosystems in general and on agriculture specifically. Growing populations require increasing amounts of food to meet even minimal needs. Uncertainties notwithstanding, many researchers have discussed the potential impacts of global warming on agriculture by way of conferences, publications, and policy recommendations. According to Rosenzweig and Hillel (1998),

> The role of carbon dioxide in agriculture is complex, in that it can be positive in some respects and negative in other respects. CO_2 concentration affects crop production directly by influencing the physiological processes of photosynthesis and transpiration; therefore it has the potential to stimulate plant growth. . . . However, many studies . . . integrating climate, crop, and market dynamics suggest strongly that the anticipated changes are likely to have large and far-reaching consequences, especially in less-developed regions (pp. 261–62).

THE TERRESTRIAL ENVIRONMENT

The media, for the most part, concentrate their coverage on highly visible, usually adverse, climate and climate-related impacts on land-based ecosystems, with prolonged droughts and floods capturing considerable attention. Often, during prolonged dry spells, premature concerns about drought fill the airwaves and the printed media, often to have those concerns evaporate when a timely rain falls and saves water-stressed field crops. With the appearance of the rains, the media's drought stories and hype come to an abrupt end.

The effects of drought on grasslands have been well documented over many decades. Continual production of annual crops in dry areas often leads to lower yields over time and can leave the exposed soils subject to wind and water erosion. The likelihood of dust storms also increases. These are forms of desertification.

In the West African Sahel, crop production is favorable during good rainfall years, but in the dry years, food shortages occur. Those who have livestock herds grazing on the seasonal grasslands will, under drought conditions, take their livestock to the relatively better watered areas to the south. This migration often leads to conflict between herders and cultivators, as the cattle and goats graze amidst the agricultural areas, destroying crops and compacting the soils. The presence of herds is favored by farmers, however, once the crops have been harvested because the livestock's manure fertilizes the soils.

Also in West Africa, farmers use flood recession farming—cultivating the floodplain and banks of rivers that had swollen during the rainy season. As river waters recede, farmers plant seeds in the exposed soils of the river's edge, where there is enough moisture to support crops. However, in the case of a hydrologic drought (a reduction in the amount of streamflow), farmers involved in flood recession farming are at high risk of crop failure and food shortages.

As the global atmosphere warms, ecosystems will shift as local climate conditions change. Tropical and subtropical climates will shift in the direction of the poles. Insects and crop diseases will follow the warmth and precipitation from the tropics into

---

BOX 4.1.

The Monarch Butterfly:
An Example of Climate-Society-Habitat Interactions

The beautiful orange and black monarch butterfly in North America has been under increasing pressure from agricultural land expansion for some decades. This butterfly winters in the forested area in Mexico's state of Michoacan, a few hours by car from Mexico City. It breeds in winter and migrates in spring throughout the United States and Canada, only to return to Mexico in the fall. The annual migratory cycle begins once again the next year.

The case of the North American monarch butterfly is interesting because it mixes national and international politics (the monarch traverses two international borders and three countries), government pronouncements as opposed to true commitments to ecological preserves, economic development conflicts in a developing country, and concern about the adverse impacts of climate change and extreme events. As a case study, it is likely to raise issues that are similar for other endangered flora and fauna, nature preserves, and biodiversity both on land and in the sea (Halpern, 2001).

Its habitat was "discovered" by American researchers in the mid-1970s, although it was likely known by local inhabitants long before then. When it was written about in *National Geographic* in the mid-1970s, one of the discoverers, butterfly biologist Lincoln Brower, later admitted that the map that accompanied the article had purposely misidentified the exact location of the forests in which the monarch was known to winter. This was done to avoid an unwanted, disruptive influx of tourists determined to see the location from which the millions of monarchs set off on their annual northward migration. An uncontrolled influx of curious tourists, however well-meaning, could destroy the monarch's habitat.

Tourists are not the only threat to the butterfly. Mexican farmers seeking to eke out a living from the land have been encroaching on this forested habitat, converting forest into farmland. Newly cleared fields have been moving persistently up the slopes (yet another example of a creeping environmental problem) in this rather small habitat of several tens of square kilometers of forested mountains. Concern had been expressed by environmental groups and later by the Mexican government about the potential extinction of the monarch as a result of habitat loss. The habitat is now an ecological preserve, but its long-term protection is not yet assured.

From the perspective of the butterfly, the climate system is yet another predator. Is the local flora and fauna resilient enough to survive a warming of the global atmosphere? If there are other changes in precipitation or temperature, will there be a corridor by which the monarch can escape to hospitable locations?

Aside from changes in long-term average climate conditions, one must also view weather and climate extremes as predators. For example, a heavy snowfall across the monarch's winter habitat in 1995–1996 killed about 10 percent of the population. Such an extreme event in the region had rarely been witnessed (Scott, 1992). Again, in February 2002, freezing rains covered the Mexican winter habitat and destroyed tens of millions of monarchs. Brower recently suggested that on this occasion "up to 80 percent of the butterflies might have died from severe weather combined with the lack of tree cover, which exposed the insects to wind, rain and cold" (Offut, 2002). Once again, questions were raised about their resilience in the face of all these natural and human predators.

the midlatitudes, thereby affecting crop production in that region.

Many organizations are deeply concerned about the negative impacts of human activities of biological resources and genetic diversity (see box 4.1). Claims to support them about the loss of biodiversity, as well as supporting evidence (terrestrial, marine, and aquatic), can be identified in all parts of the globe and on all types of ecosystems.

Biodiversity is shorthand for biological diversity. It is defined as "all the variety of life that can be found on earth (plants, animals, fungi, and microorganisms) as well as to the communities that they form and the habitats in which they live" (Belgian Clearing-House Mechanism, 2001). It can be affected by geophysical as well as human factors. The specter of global warming poses an additional threat to existing degrees of biodiversity. "Climate change may directly affect species through changes in phenology (e.g., earlier flowering of trees and egg-laying in birds), lengthening of the growing season, and changes in distribution, resulting from migration (e.g., poleward and altitudinal shifts in insect ranges). In many cases the observed changes are consistent with well-known biological responses to climate" (CBD, 2002).

Conservation International has identified more than 20 locations as biodiversity hotspots, or areas at risk to species extinction resulting from human activities and global warming (Conservation International, 2002).

"Ecological forecasting," a new activity of the U.S. government, is designed to monitor changes in U.S. ecosystems, both managed and unmanaged. Its creation was based on the belief that "to sustain our valuable ecosystem goods and services, we have to understand how ecosystems function and interact and, more importantly, forecast how they will be affected by change" (CENR, 2001, p. 1). Many of the ecological changes of concern to the government are climate-related (see figure 4.6).

THE MARINE ENVIRONMENT

The "marine environment" is a misnomer. There are many marine environments: estuaries, the coastal ocean, the high seas, deep water, shallow shelves, bays, deltas, regional seas, and so forth. Each of these environments provides a hospitable habitat for certain

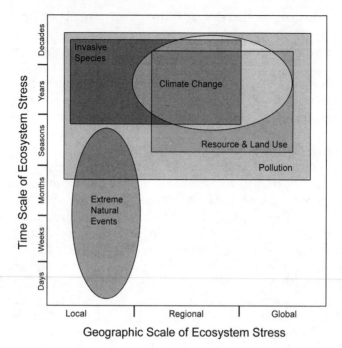

Figure 4.6. Five key causes of ecological change play out and interact on a wide range of time and space scales. Even this oversimplified graph suggests that one of the greatest forecast challenges will be to predict the cumulative impacts of multiple stresses. (Source: CENR, 2001, p. 1)

types of living marine resources. These living marine resources have adjusted over time to the normal regional or local climatic variations that occur on various time scales from seasons to decades. However, with a global warming of the atmosphere and of the oceans, local and regional climatic norms would change. This, in turn, would disturb the local and regional habitats to which the different living marine resources have become accustomed.

The interactions between the ocean and the atmosphere are multifaceted in time and space. Perhaps El Niño is one of the best examples because it is the result of interactions between oceanic and atmospheric processes in the tropical Pacific. Another example would be the location, direction, and strength of major ocean currents such as the Gulf Stream in the Atlantic and the Kuroshio Current in the Pacific. Yet another example is the way that carbon is taken out of the atmosphere by oceans and sequestered in the deep ocean for centuries. The impacts of changes in the atmosphere on marine ecosystems and on living marine resources are highly complicated, as can be seen in figure 4.7 (Glantz, 1992). The variability and change of local or global climate can foreseeably affect each stage in the life cycles of marine plants and animals.

Climate impacts on marine, as opposed to terrestrial, ecosystems have received less attention for a variety of reasons:

- most fishing activities do not take place where people live;
- the fishing sectors in most countries involve a relatively small number of workers;
- it is mostly an economic problem in the industrialized countries, whereas it can be one of life and death in artisanal (small scale, often subsistence) fisheries in developing countries; and
- the relationship between the viability of living marine resources and variability in air-sea interactions in different parts of the world's oceans are much more difficult to study and are not so well understood.

There are notable examples, however, of detailed studies of climate-marine-society interactions: the California sardine, Peruvian anchoveta, Ecuadorian shrimp, Pacific Northwest and Alaskan salmon and Alaskan king crab, and Atlantic cod, among others.

The executive summary of the first Intergovernmental Panel on Climate Change (IPCC) (1990) scientific assessment suggested how climate change might affect marine ecosystems and living marine resources. It noted:

Climate change is one of the most important factors affecting fisheries. The level of impact varies widely and depends on attributes of the species as well as on their regional specificity. . . . Changes in ocean circulation may lead to the loss of certain populations or the establishment of new ones. . . . Warming impacts on the abundance of

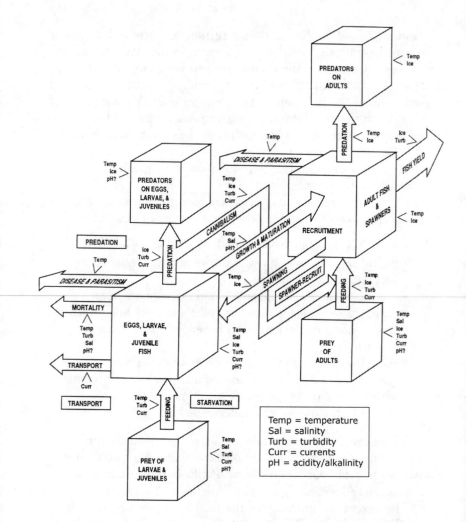

Figure 4.7. Major biotic processes affecting fish production and the abiotic factors that modify these processes. The four major hypotheses concerning control of fishery abundance are related to the major processes controlling production and mortality of early life history stages: reproductive output, starvation, predatory (including cannibalistic) losses, and transport losses. To represent an actual fishery environment, several such interlocking diagrams would be needed to depict multiple species. (Source: Glantz, 1992, p. 6)

commercially important species can be either negative or positive, even for the same species, depending on the region. . . . It is likely that global warming will produce collapses of some fisheries and expansion of others. (as quoted in Glantz, 1994b)

As the global climate warms, so too will the oceans, suggesting that there will likely be a poleward shift of the area inhabited by warmwater and coldwater species. These shifting locations of various living marine resources will have an obvious second-order impact on economies and communities presently dependent on their exploitation. There are already some examples that merit study of societal adjustments that have been made in the face of a loss of access to fish stocks (e.g., impact in Japan of the Hokkaido herring collapse in the early 1900s, the impact in Great Britain of the loss of access to Icelandic cod in the mid-1970s, the collapse of the cod fishery on the Georges Bank in the 1990s, impact on the U.S. and Canada of the decline of salmon in the Northeast Pacific).

Coral reefs have increasingly been in the headlines for the past decade because of worldwide coral bleaching. Attention to the bleaching process increased sharply as a result of the 1982–1983 El Niño event (Glynn, 1990). Researchers noted that coral bleaching was taking place because of a significant increase in sea surface temperatures. Since then, coral reefs have been closely monitored for changes that might be linked to El Niño-related or long-term climate change-related ocean warming.

The present description of climate impacts on ecosystems attempts to separate natural processes and human activities. However, these processes interact. So, in the marine environment, living marine organisms are subject to variability, fluctuation, and changes in the oceanic and atmospheric environments as well as to exploitation by insatiable predators, such as fishing industries. The combination of such pressures on living marine resources is a determining factor in their potential for long-term sustainability of what are seemingly renewable resources.

Impacts on Society

Most people are aware of the adversities associated with droughts, floods, freezes, and fires, whether these events have directly affected them or they have learned from the media that these events have occurred elsewhere. Impacts ripple through society and often across international borders. For example, drought in the American Midwest adversely affects crop production, and that will in turn affect farmers' ability to buy new farm equipment, pay taxes, or put new land under irrigation. It also affects commodity prices in the market and the availability of grain for humanitarian food aid around the globe.

The impacts of climate and climate-related anomalies are of great concern to the general public everywhere and to their governments. They can adversely or positively affect food production and prices, water resource availability, energy production and consumption, fishery abundance, public health, and public welfare in general. For example, although a drought may last for a few months, its impacts can linger for years. The same is true for the impacts of other extreme climate hazards.

Commenting about rangeland overgrazing in the western United States several decades ago, Dasmann (1959) wrote that

> In the early days . . . stockmen from the East, inheritors of European traditions, were familiar with livestock management on the well-watered pasturelands of these regions. They had no experience with the arid lands of the West, where the capacity of the land to support livestock is often extremely low (p. 191).

Dasmann further commented that "climate has been, and remains, a major cause of range damage, although it is often blamed for man's mistakes" (p. 192). Such misdirected blame continues today.

The severity of the impacts on society and ecosystems of climate and climate-related anomalies is not simply the result of adverse climatic conditions but is also a function of the level of vulnerability of a society. The time for recovery from the impacts of a drought, tropical cyclone, or El Niño-related bush or forest fire will depend on the level of resilience of a society. Resilience refers to the capability of a society to rebound from an adverse climate-related impact. For example, wealthier societies have the economic resources to cope with the impacts of an extreme event and to rebound rather quickly from those impacts.

Whether a particular government is resilient enough to withstand such an extreme is, however, another matter. The impacts of an extreme event in the same location at different times will likely vary depending on what is going on in that society at that particular time, politically, economically, and culturally.

SEASONALITY

To date, interest in climate impacts on societies and on rural poverty has focused on spectacular, unusual, or extreme events such as droughts and floods. However, such events often distort socioeconomic relationships that have already been established by the natural rhythm of the seasons, or seasonality. This is something to which every society has had to adjust. That flow, however, is often disrupted by extreme climatic events, and how societies cope with those disruptions will determine the success or failure of human activities. This disruption has had major implications for rural poverty in the Third World (e.g., Sahn, 1989).

A large portion of the world's population is made up of subsistence farm families that live from harvest to harvest. Anything that disrupts the expected flow of the growing season, from field preparation to harvesting, or the ability to exploit that flow, disrupts their ability to grow enough food for themselves and for sale or trade in the local markets. In this regard, well-off farmers try to store enough food to survive a bad crop season. Some of the most well-off (relatively speaking) are able to cope with two bad crop years in a row because they have on-farm and off-farm strategies and tactics to survive the impacts of seasonal or interannual climate variability. Few farmers, however, can withstand prolonged droughts of more than a couple years without getting outside help or, in developing countries, without having to flee their land to the slums of urban centers or refugee camps.

Chambers et al. (1979) summarized how seasonality relates to rural poverty. They noted that "besides climate, seasonal patterns are also found in labor demand in agriculture and pastoralism, in vital events, in migration, in energy balance, in nutrition, in tropical diseases, in the condition of women and children, in the economics of agriculture and in social relations, and in government interventions" (p. 3).

The seasonal rhythm tends to reinforce the existing social and economic relationships between the local "haves" and "have-nots," while extreme meteorological events such as drought tend to exacerbate an already distorted socioeconomic relationship between them. Thus, understanding seasonality and the relationships of drought and seasonality to rural poverty is essential for establishing effective ways to mitigate climate-related societal impacts.

On the other hand, knowing how seasonality and its variations can benefit or harm human activities and societal processes, and then using that information to enhance benefit and minimize harm, is an excellent example of exploiting existing climate as a resource.

Speculation notwithstanding, it is not yet clear how seasonality, as witnessed in different parts of the world, will be affected by global warming. Although the average temperature of the atmosphere may increase, this does not mean that all temperatures will increase at all times. Summers may become longer and winters warmer. Summers may become wetter or drier, depending on location. We may even witness hotter summers and colder winters in some areas. In the Aral Sea basin, for example, a decline in the sea's surface area and volume since 1960 has been accompanied by a change in the region's climate, which has become more continental—that is, hotter summers and colder winters. Seasonality is so important to societies everywhere that it demands more attention, not only from researchers in the physical and social sciences but from policy makers and their economic planners as well.

WHO IS MORE VULNERABLE?

Speculation abounds about whether rich countries or poor ones are more vulnerable to climate variability, change, and extremes. Although people in rich industrialized countries seemingly have the economic wherewithal to react (if not pro-act) to impacts, in many respects they exhibit a lower level of tolerance for discomfort than those in poorer countries. Perhaps this is because inhabitants of developing countries have had more experience in coping with and accepting the adverse impacts of climate anomalies than those in industrialized countries. Their social structures and cultural practices have developed ways to assist those who have been adversely affected by a climate-related hazard. For example, pastoralists in many parts of the globe have developed kinship ties with farmers in areas with reliable rainfall and agricultural production. In times of drought-related stress, they rely on those kinship ties to get them through. The issue of comparative vulnerability and resilience among societies at different levels of economic development remains an important research question.

Climate Politics

Politics is said to exist wherever issues arise. Climate politics refers to the varied processes pursued by different actors to achieve a policy objective desired by one group or nation, often at the expense of the objectives of other groups or nations. Many climate-related political issues are played out at local, national, regional, and global levels and are in need of attention and resolution. In addition to the traditional political actors responsible for making climate-related policies and laws, such as diplomats, local and national politicians, scientists, and bureaucrats in a variety of government agencies, one needs to add nongovernmental organizations—lobby groups, environmental, religious, and other special-interest groups—to the mix.

For centuries, governments, acting alone or in combination with others, have made regulations, laws, policies, and nonbinding resolutions relating to naturally occurring climate extremes. They have also sought to manage, if not influence, climate variability and extremes and their impacts through irrigation development, tree planting (shelterbelts), controlling flood plain development, restricting the use of grasslands, devising ways to share fish stocks, and so forth. For example, in medieval times, rulers banned settlements near the Caspian Sea's shore on the threat of death, based on their knowledge of the dangers to structures of fluctuating levels of the Caspian Sea.

In the late 1960s, climate issues were starting to raise the eyebrows of political leaders in some countries. Until then, studying climate, as far as students and economic development experts were concerned, was seen as equivalent to studying

Latin—interesting perhaps, but not very useful. It seemed that climate studies had been relegated for the most part to historians, geologists, geographers, archaeologists, and climatologists. Climate considerations were also of interest to some researchers in disciplines whose operational activities climate affected: human settlements, forestry, agriculture, water resources, energy, and health.

Specific political interests in the late 1960s about the possible adverse impacts on the stratospheric ozone layer of a large fleet of high-flying supersonic transports was symbolized by the British-French joint venture called the Concorde. Although President Nixon supported the development of a U.S. fleet, scientists and an emerging environmental community opposed it. The final report on the issue, which was the first blockbuster study of atmosphere-society interactions, concluded that chemical emissions (nitrous oxide) from a large fleet of these aircraft flying in the stratosphere would deplete the protective ozone layer, although its executive summary reflected a more positive perspective (Grobecker et al., 1974). This led to congressional hearings and a subsequent ban on Concorde landings at U.S. airports for some time.

Political interest in climate-society interactions sharply escalated in the early 1970s as a result of the truly surprising occurrence of a cluster of notable climate anomalies. An El Niño–related collapse of the Peruvian fisheries in 1972 and 1973 had a major impact on the global food crisis. Severe droughts and water shortages in the late 1960s and early 1970s set the stage for the convening of several UN conferences related to world food insecurity and to the environment. Numerous droughts in the early 1970s were linked to the 1972–1973 El Niño.

Also at that time, the Soviet Union's grain-producing "bread-basket," the Ukraine, suffered tremendous climate-related grain production losses. In response, the Soviet government was forced for the first time in its history to buy large quantities of grains in the international marketplace. The crop losses were due in part to drought but also to poor agricultural practices on the government's collective and state farms. The second-order effects of the Soviet crop failures were major. Massive Soviet imports of corn and wheat from the United States, referred to as the Great Grain Robbery, worsened the drought-related food shortages that existed in many developing countries at the time (Trager, 1975). In addition, large increases occurred in the prices of various grains in the marketplace. This collection of anomalies prompted *Fortune Magazine* to commission an article entitled, "Ominous Change in the Weather?" (Alexander, 1974).

Dwindling food stocks in grain-exporting countries (e.g., the United States, Canada, Australia, Argentina) prompted the United Nations (UN) to convene the World Food Conference in 1974 (Brown and Eckholm, 1974). The UN also convened a World Population Conference in the same year. A series of other UN-sponsored international meetings followed throughout the rest of the 1970s on water, human settlements, desertification, technology, and climate.

In 1979, the UN's World Meteorological Organization (WMO) convened the first World Climate Conference in Geneva, Switzerland. This led to the creation of the WMO's World Climate Program, along with several national climate programs, such as the U.S. National Climate Program Office. Although this U.S. office no longer exists, it was instrumental in generating national and global interest and research on climate-related issues, including aircraft noise pollution; nuclear tests; spray can aerosols (chlorofluorocarbons [CFCs]); impacts on stratospheric ozone; global cooling (yes, global cooling); and traditional agriculture; and energy and water resource concerns. Cloud seeding for the purpose of weakening the damaging effects of hailstorms and for reducing hurricane intensity (e.g., Project Stormfury) also became popular but politically contentious research areas. The fear that scientific research related to climate modification could become a weapon of war was also raised. In fact, fearing such a prospect, the U.S. Senate passed a resolution to ban the use of the atmosphere for purposes of war (U.S. Senate, 1973).

Today, climate and climate-related politics have come of age. They receive attention from the general public, the media, politicians, and forecasters. Climate and climate-related policies and laws encompass a wide range of issues from air pollution and acid rain to global warming and ozone depletion and include subnational, national, and transboundary issues related to hydrologic and atmospheric processes. Library shelves are now filled with books, journal articles, and unpublished literature, for example, government reports, on these and many other climate-related law and policy issues.

An interesting recent example of a transboundary climate-related problem involves the Pacific salmon fisheries. Salmon are anadromous, meaning that they are hatched in freshwater, migrate a few thousand miles in the salty water of the open ocean, and return after several years to their native streams to lay eggs in freshwater. On their migratory paths, salmon are vulnerable to fishermen of many countries in the open ocean and in their native streams as they traverse the territorial waters of neighboring countries, states, and provinces.

For several decades, the United States and Canada have had a treaty about the sharing of catches of salmon in the Pacific Northwest. The Fraser River catches before 1975 were about equal between the continental United States (Pacific Northwest states) and Canada (British Columbia). Political cooperation and understanding between the United States and Canada on this issue seemed to break down after the 1976–1977 El Niño, when warm waters in the eastern Pacific altered the productivity and migratory patterns of the salmon. There were dramatic increases in Alaskan salmon stocks, and the Fraser stocks returned via the north end of Vancouver Island, thus avoiding U.S. waters. This migratory shift contributed to the salmon depletion along the Washington and Oregon coasts. Canadian fishermen were able to capture almost 90 percent of the Fraser salmon landings, as opposed to the agreed-upon 50-50 split, while Alaskan harvests of Canadian salmon also increased. This altered situation continued for the next fifteen years. The changes in salmon abundance and harvesting opportunities, which were linked to the shift that had apparently occurred in the ocean environment after the mid-1970s, led to two periods of intense conflict. The first was resolved by the 1985 Pacific Salmon Treaty, but that agreement quickly broke down. The end result, after much contentious debate punctuated with hostile actions by both parties, was the signing of a revised agreement under the Pacific Salmon Treaty in June 1999 (Miller et al., 2001).

This situation in some ways was not unlike the attempts in the early 1920s to divide the streamflow of the Colorado River between upper and lower basin states in the United States. In both situations, agreements were made among legal entities to share resources based on a belief that the environmental conditions in the future would be as they were in the relatively recent past. In both cases, those assumptions were proven wrong because of environmental changes on the decadal time scale. In fact, with regard to the salmon issue, researchers have identified four such migration fluctuation periods in the twentieth century (see, for example, Taylor and Southards, 1997; see also Miller et al., 2001).

Nongovernmental Organizations and Climate Politics

Nongovernmental organizations (NGOs) have been deeply concerned about a wide range of aspects related to environment-society interactions. They have become increasingly involved in climate and climate-related issues. For example, it is not possible to talk of biodiversity issues without bringing climate into the discussion. The same is true for food, energy, water, health, and public safety concerns. Even local grassroots NGOs centered on recycling, saving a specific wetland, planting trees, or cleaning up a river or one of the Great Lakes could not help but become engaged in climate-related discussions.

Representatives of the larger national and international environmental NGOs (such as the Sierra Club, the Environmental Defense Fund, Greenpeace, and Conservation International, among many others from around the globe), private voluntary organizations, and humanitarian organizations have often sat alongside government officials at conferences for both national and international policy-related deliberations on climate issues. This is especially true for climate change issues. Their influence on the direction and content of negotiations toward environmental protection is noticeable. Groups that have tended to challenge governments' policies have participated in parallel conferences held at the same time and in the same location as, for example, the Earth Summit in Rio, the Conferences of Parties meetings, and other major climate-related meetings. In many instances, members of environmental groups have taken their political protests to the streets over the lack of government support for reducing GHG emissions. This was obviously the case during President Bush's visits to Japan, China, and South Korea in mid-February 2002.

What is the real value of NGO involvement in climate and other environmental negotiations? Some suggest that their influence may not be very strong, but that they should be encouraged by governments to participate. Some NGOs and scientists have better expertise than government diplomatic negotiators. The NGO perspective might shed new light on issues under discussion. Perhaps most importantly, NGOs often bring transparency to the negotiating process, keeping governments and the results of their negotiations more honest and, therefore, acceptable to citizens (Chasek, 2001, p. 231).

It is important to note, however, that all NGOs are not equal, nor are they treated equally. Greenpeace, for example, is viewed as much more radical in its tactics than the larger, more accepted NGOs such as the Sierra Club, Environmental Defense Fund, World Wildlife Fund, Conservation International, and World Resources Institute. There is also a gap in influence among the NGOs; the smaller grassroots organizations are left outside the decisionmaking or policy-influencing circles, hence the strong need for parallel unofficial conferences.

A 1993 UN High Commissioner for Refugees report highlighted some general contributions of NGOs to political activities:

> Modern NGOs seek to influence the policy decisions of the international community by sensitizing public opinion, the news media and politicians to important issues, by recommending actions and by exposing failures and abuses. . . . Many NGOs can operate unencumbered by the political constraints which sometimes hamper the policies and actions of inter-governmental and national governments, taking as their constituency those who are least able to present their own cases. (UNHCR, 1993, p. 128)

Clearly, there are linkages between climate policy and climate politics. Political debates, conflicts, and shifting coalitions are the political pathways that support or block the passage of climate policy and law. In other words, policy and law are what you get; politics is how you get there. Examples of some of the political tensions involved in climate politics can be thought of in shorthand: the White House effect vs. the greenhouse effect, conservation vs. exploitation, environmental movement vs. wise use, and technophiles vs. technophobes. These tensions are explored in the following sections.

The White House Effect vs. the Greenhouse Effect

Several years ago, an article appeared in a politically conservative, influential Washington D.C. newspaper, the *Washington Times* (Brookes, 1990), headlined "The White House effect vs. the greenhouse effect." It captured the heart of political, if not ideological, debate over climate change. The article's headline challenged the science of the global warming issue, attacking it as being a key part of the environmentalists' political agenda to shift America from an energy dependence on fossil fuels to renewable energy sources. The energy sector and conservative members of Congress led the attack. Al Gore's 1992 proenvironmental movement book *Earth in the Balance* helped to fuel this ideological divide.

In the U.S. presidential election of 2000, Governor George W. Bush of Texas defeated Vice President Gore. The Bush White House is much more sympathetic than Clinton's to the energy industry's concerns about restricting the burning of fossil fuels. As a direct result of the shift from a proenvironment administration to one in favor of support for the energy industry, the U.S. global warming policy was reversed.

Under the George W. Bush presidency, the phrase "the White House effect vs. the greenhouse effect" has taken on exactly the opposite meaning from the domestic political conflict as was captured in its original use under a Republican president. Although the human contribution to the naturally occurring greenhouse effect continues to grow, the Bush administration offers diametrically opposed views and policies about global warming and what to do about it in terms of domestic energy policy from those of the Clinton administration a few years ago.

Most recently, California's political leaders passed the United States' first law to pursue a zero-emissions policy for vehicles. This was a move by Californians to reduce their greenhouse gas emissions in response to the national government's inaction on global warming. However, as noted in the *Wall Street Journal* (Ball, 2002), "the Bush administration's Justice Department submitted a legal brief supporting General Motors Corporation and Daimler Chrysler AG's Chrysler unit in a lawsuit . . . filed against a California clean air effort." They claimed that the California program was illegal "because it amounts to a state effort to regulate automotive fuel economy, an authority reserved for Washington under federal law" (p. A-2).

Conservation vs. Exploitation

Conservation and exploitation represent two opposing perspectives on the environment and especially on the use of natural resources. Perhaps there is no better example of these two opposing views than the perspectives of the two candidates in the 2000 U.S. presidential election. Vice President Al Gore represented most conservationists. His policies sought a balance between resource use and conservation of the environment. Had he won the election, his U.S. Environmental Protection Agency (EPA) would likely have sought tighter controls on urban air pollution, acid rain, and the emission of carbon dioxide as a result of fossil fuel burning. Bush, on the other hand, favors using natural resources including fossil fuels to allow Americans to maintain their current lifestyles; easing restrictions on urban air pollution; pulling out of the Kyoto Protocol, an international agreement to reduce national GHG emissions to 1990 levels within a decade; opening wildlife reserves, public lands, and coastal areas to oil and gas exploration and transport in the name of energy independence; and allowing commercial logging a free hand in the nation's forests.

The president of the NGO Friends of the Earth put forth the following observation about Bush's environmental policies: "In my more than 30 years of working to protect our planet, I have never seen an administration more eager to take a stand against virtually every environmental law and protection ever written

or enacted. . . . Friends of the Earth is convinced that the administration has been using the cover of terrorism to dismantle decades of progress on basic environmental protections" (Blackwelder, 2002).

The tropical rainforest in the Brazilian Amazon is one of many other examples of a battleground, so to speak, between those who want to exploit natural resources and those who want to preserve them. Deforestation has resulted from tropical timber exports, land clearing for agricultural purposes, influxes of migrants from drought-afflicted Northeast Brazil, road construction (the Trans-Amazon Highway), the search for gold, and government economic development policies.

Environmental Movement vs. "Wise-Use" Movement

"Wise use" is a movement that calls for the extensive use of natural resources; it is basically a "use it or lose it" philosophy. Its supporters believe in economic growth first and foremost, and its views are opposite of those expressed and pursued by environmental groups in general and especially those expressed by the more radical Greenpeace, a proactive organization with an in-your-face approach to making its views known. Wise-use representatives have compared their views with those of the environmental movement. They noted, for example, that many environmentalists believe that "we all live downstream," meaning that eventually we will all suffer the consequences of environmental degradation. Wise-use people, on the other hand, argue that "we all live upstream," meaning that the earth is bountiful and that Wise Users are its stewards.

The group Public Employees for Environmental Responsibility (PEER) succinctly presents a different view of the Wise-use movement.

> Groups calling themselves members of the "Wise Use" movement represent landowners, loggers, off-road vehicle users, miners and other individuals who are often at odds with legislation enacted to protect and preserve public lands. The term "Wise Use" was co-opted from Gifford Pinchot, first chief of the Forest Service, who advocated

the wise use of natural resources. Unfortunately, anti-environmental groups are committed to opening up public lands not only to harmful resource extraction, but to outright exploitation. (PEER, 2001)

There appears to be little chance for any meaningful dialogue between those who hold these conflicting views of what the appropriate society-nature relationship ought to be.

Technophiles vs. Technophobes

Technophiles believe that science and engineering can be used to overcome obstacles to addressing climate and climate-related problems. They believe that societies can dominate nature and that natural resources are there to be used. They feel that if environmental problems arise, scientists and engineers can be called on to find ways to resolve those problems.

For example, the Soviet leaders Lenin, Stalin, and Khrushchev, among others, supported major environmental changes in their drive to dominate nature for purposes of large-scale economic development: the Virgin Lands scheme to convert grasslands into large collective farms in the 1960s and 1970s; the construction of the longest human-made Karakum Canal in Turkmenistan after 1954; and Siberian river diversions to arid Central Asia, to name a few. Other societies as well have held similar beliefs and have undertaken similar large-scale engineering feats in the name of economic development (e.g., Farvar and Milton, 1972; Kluckhohn and Strodbeck, 1959).

The U.S. Army Corps of Engineers has spent more than a century and a half devising engineering ways to channel waters in the mighty Mississippi River and its tributaries, only to have natural forces overwhelm the manmade levee system in 1993, causing the river to recapture large segments of its natural floodplain during what became known as the Great Midwest Flood of 1993 (Changnon, 1996). In every country, one is likely to identify several examples of the conflict between those with a blind faith in the use of technology and those who believe that the use of technology to cope with climate-related problems is often as harmful as it is beneficial.

Technophobes favor the use of "appropriate" technologies to

minimize the impacts of human activities on the environment and to maximize the sustainability of human-nature interactions (Schumacher, 1989). Technophobes do not share the technophiles' blind faith in technological fixes. Nor do they trust their country's scientific and engineering establishments to consider all of the actual environmental and social costs of new technologies that are constantly being developed to address societal needs. They do not trust representatives of governments or corporations because they believe that these representatives put personal political gains or corporate profits ahead of public welfare or environmental protection.

Where National and International Politics Meet

It was a strong hope of scores of governments to seek ratification of the Kyoto Protocol by August 2002 in time for the tenth anniversary conference of the 1992 Rio Earth Summit. However, as noted earlier, the U.S. presidential election of 2000 resulted in a radical shift in environmental philosophy in the White House and, more specifically, in national climate change policy. After a couple of months in office, Bush boldly announced his plan to open the Alaska National Wildlife Refuge to oil exploration, a region that many American people, especially environmental groups, have tended to view as off-limits to the oil industry.

Bush finally announced his new U.S. policy toward GHG reduction in mid-February 2002, and it was based on corporate volunteerism. It entailed corporate actions to reduce greenhouse gas emissions and proposed various economic tax favors to the corporations to encourage corporate involvement. In Bush's words, "My approach recognizes that economic growth is the solution, not the problem. Let's do what is in the interest of the American people" (Revkin, 2002). In his speech, the president said that his goal is to slow down the rate of increase in greenhouse gases and to reduce emissions from power plants of air pollutants such as sulfur dioxide, nitrogen oxides, and mercury. Neither activity would help to curb global warming. A German climate change specialist (H.-J. Schellnhuber) referred to his policy as "a wonderful policy, but the wrong target" (Daley, 2002).

The United States was the only government to withdraw from the Kyoto Protocol negotiating process. It was clear who would support and who would attack Bush's new global warming policy, both domestically and internationally. Although several corporations supported his go-it-alone policy on climate change, governments and environmental groups attacked it as a do-nothing policy. European allies attacked the Bush policy, expressing a fear that it could undermine the fragile international cooperation and progress made at the Conference of Parties in Morocco in late 2001.

In mid-February 2002, Canada's nine provincial prime ministers surprised the federal Prime Minister, Jean Chrétien, when they announced their opposition to Canadian support for the Kyoto Protocol. Till then, Chrétien had no idea that the provincial leaders had been meeting on the topic and was visibly surprised and annoyed when they made their announcement at a public press conference. It appears that Bush's policy of voluntary approaches to GHG reduction had hidden supporters. "The premiers believe the 'rigid and stringent' Kyoto agreement puts Canadian business at an unfair disadvantage because neither the United States nor Mexico will be covered by the environmental treaty," said Alberta's Premier Klein (Harper, 2002). Nevertheless, Chrétien's politicking to ratify the Kyoto Protocol won the day by December 2002, further isolating Bush's position on climate change (Ljunggren, 2002).

Another example of the interplay between national and international politics came when an Indian expert in economics and technology, R. K. Pachauri, replaced the chairman of the IPCC scientific assessments for climate change, Robert Watson, in mid-April 2002 as a result of U.S. government pressure. Pachauri will lead development of the Fourth Assessment Report, due in 2007. The reason for Watson's dismissal was placed squarely on the influence of energy corporations in the Bush administration. Watson, the head of the World Bank's Environment Department, was viewed as a passionate advocate of the need to reduce greenhouse gases by reducing national dependence on fossil fuel use. Pachauri, the director of the TATA Energy Research Institute (TERI), was seen as being more objective, at least by those who opposed a chairman who was an advocate for action to stop global warming (Fialka, 2002).

The politics of global warming are going on around the globe as societies grapple with global warming science, scenarios for the future, energy needs, biases, and more broadly, with competing perspectives on society's dealings with the atmosphere in general. California's recent passage of legislation focused on taxing in-state greenhouse gas emissions to reduce the state's emission levels indicates that global warming politics are not simply an international political issue.

Climate-Related Aspects of International Security

Burroughs (1997) made the following observation:

> The subject of the historical impact of climatic change stirs emotions. There is a range between those who argue that fluctuations in the climate were wholly inconsequential in the course of history to those who believe they are an unrecognized mainspring which has controlled the outcome of many events. As with so many fiercely debated issues, reality is somewhere between these extremes (pp. 1–2).

Anyone who has taken a European history course has learned about the failure of Napoleon's invasion of the Russian Empire in 1812. This is truly one of history's worst military defeats, and it was not necessarily brought about by an opposing army alone, but by anomalously harsh winter-like conditions.

Figure 4.8 depicts the timeline of Napoleon's invasion of Moscow along with the temperatures during his retreat. As they entered the Russian Empire to begin their march on Moscow, Napoleon's troops numbered in the hundreds of thousands. However, Napoleon's decision to try to capture Moscow at the onset of the cold season planted the seeds of his downfall. The weather patterns were considerably wetter and colder than normal, and these winter conditions occurred much earlier than usual. Of the several hundred thousand troops that began the invasion, less than a few tens of thousands made it back to their starting point. Napoleon was purported to have remarked, "I came to fight men, not nature."

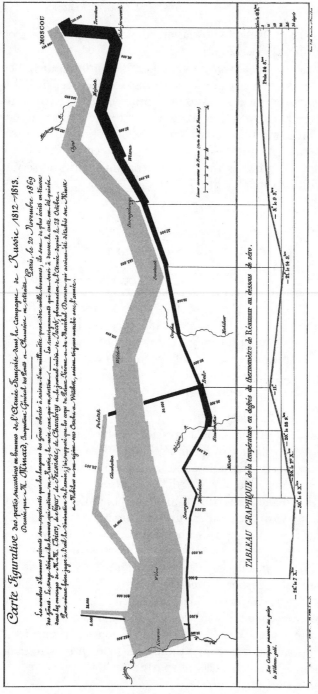

Figure 4.8. Napoleon's invasion of Russia, 1812; based on the 1869 chart by Minard. The shaded area varies in width to show how the invasion and cold temperatures took their toll on troop mortality during its march into Russia and its retreat to Poland. Light-shaded area shows troops en route to Moscow; dark-shaded area shows their retreat. The bitterly cold temperatures became Napoleon's worst enemy during this campaign. (Source: Edward R. Tufte, *The Visual Display of Quantitative Information*, Graphics Press, 1983)

Figure 4.8 highlights the often hidden influence, positive or negative, that weather or climate extremes can have on the course of history and on a government's security. Napoleon's march on Moscow is but one such example of how variations in seasonal climate have affected, if not altered, the course of history. More than a century after Napoleon's defeat, Hitler's armies invaded the Soviet Union during wintertime. The Nazi Army suffered a similar humiliating defeat at the hands of Russian partisans and their natural ally, a severe Russian winter that was one of the coldest of the twentieth century.

An interesting, little-known military incident took place in Europe in winter 1794–95 (figure 4.9). Napoleon's troops were at war with England and its allies and had been advancing on what is now The Netherlands. That winter was particularly wet and cold, with rivers and harbors freezing over. As a result, part of the Dutch fleet was frozen in at the port of Texel. One of Napoleon's officers foresaw an opportunity to take advantage of

Figure 4.9. Painting by Charles Louis Mozin (1806–1862). The French Cavalry led by General Deynter take the battle fleet caught in the ice in the Port of Helder to the waters of Texel, 21 January 1795. (Reprinted with special permission, Réunion des Musées Nationaux/ Art Resource, New York)

the cold and icy conditions and ordered his horsemen onto the harbor's frozen waters to capture the icebound Dutch fleet. It was indeed a rare, if not unique, occasion when men on horseback could capture several ships at anchor. French General Jomini (1820) wrote a military account of this incident:

> A new feat of ingenuity was about to draw attention to this already remarkable campaign. Pichegru had sent an advance party of cavalry and light artillery to North Holland, with orders to cross the Texel, and approach and capture the Dutch war vessels known to be anchored there. It was the first attempt at capturing a fleet on horseback. And the attempt succeeded well beyond all expectations: the French galloped across the frozen waters toward the vessels, where they ordered the Dutch to surrender and captured the naval fleet without a struggle (Vol. 6, Book 7, Chap. 152, p. 208, original in French).

Contemporary Examples of Climate, Weather, and the Military

There are plenty of examples of weather and climate influences on military tactics and strategies, not only a century or more ago, but in recent times as well. One notable example of the importance of a forecast relates to forecasting the Allied invasion at Normandy in June 1944. Cox (2002) described the situation in the following way: "Two million men were waiting on six. In a state of what their commander called 'suspension animation,' combat troops intent on liberating Europe from the armies of Adolph Hitler were waiting for a group of weathermen to decide if conditions would permit the largest military invasion in history to go forward on June 5, 1944—or not" (p. 189). He quoted General Eisenhower who had noted in his book, *Crusade in Europe*, "the selection of the actual day would depend on the weather" (Cox, quoted on p. 190). Better weather conditions were forecast for June 6, and the rest is history.

In the early 1960s, the U.S. government supported a cloud-seeding effort in the tropical Atlantic. Called Stormfury, this was an experimental research program to reduce the strength of potentially destructive hurricanes and tropical storms with silver iodide. Hurricane modification had been proposed to reduce the costs of impacts of tropical storms and hurricanes in the region. Questions about the early successes of hurricane modification, along with growing political opposition in Central America to cloud seeding of hurricanes, brought a halt to this activity in 1980. Those opposed to cloud seeding claimed that the operations were for military purposes and also that the rainfall associated with the hurricane season was needed each year for food production, water resources, and hydropower throughout the region.

During the U.S. war in Vietnam in the late 1960s and early 1970s, the U.S. military resorted to cloud-seeding activities along the Ho Chi Minh Trail, a supply route through Laos and Cambodia used by the North Vietnamese to support their war effort and Viet Cong allies in South Vietnam (Shapley, 1974). Although the science of cloud seeding was then, as it is now, surrounded by significant uncertainties as well as scientific controversy, the military had hoped, through Project Popeye, which ran from 1966–1972, to prolong the monsoon season, making the Ho Chi Minh Trail muddier and more impassable to the Viet Cong. It was a failed attempt to slow down the shipment of military supplies and troops to South Vietnam from the north.

In early November 1979, the government of the Shah of Iran was overthrown by the supporters of the Ayatollah Khomeni. Americans were taken hostage at the U.S. embassy in Iran. As negotiations for their release failed, the U.S. president (Carter) ordered a clandestine military operation (Eagle Claw) in April 1980 to free the embassy hostages in Tehran. The rescue operation failed in part because the U.S. military planners apparently did not take into account the impact of seasonal dust storms, a known condition of the region's arid climate, on the ability of helicopters to function. Three of the eight helicopters were damaged in a sandstorm (Columbia Encyclopedia, 2000).

By the mid-1980s, considerable scientific research and political attention focused on the potential impact on the global atmosphere of a nuclear war (e.g., Ehrlich et al., 1984). That attention produced several scenarios, called "nuclear winter," that took into account the adverse impacts on the atmosphere of nuclear warhead explosions of varying megatonnage. Researchers identified the real possibility of a rapid cooling of the earth's atmosphere because of an increase in the opaqueness of the atmosphere that would result from airborne debris caused by the explosions. This would lead to the likely inability of sufficient life-supporting solar radiation to reach and heat up the earth's surface for some years. United States and Soviet researchers worked separately as well as collaboratively on this topic of global concern (e.g., Turco et al., 1984).

In 1942, Peru defeated Ecuador in a border war, the results of which Ecuador did not accept. The conflict between these countries re-ignited in 1995 over an undelineated border area that allegedly contained gold, uranium, and oil. Both countries mobilized their navies, armies, and air forces as the war heated up. With the onset of El Niño in 1997, heavy rains and thick cloud cover were expected in the region of the disputed border. *El Universo,* an Ecuadorian newspaper, reported in October 1997 that the hot war would be difficult to carry out, because the planes and tanks could not be used under difficult weather conditions associated with El Niño events. A peace treaty was signed in late October.

Today, military establishments such as the U.S. Navy and the U.S. Army maintain websites about, for example, the influence of El Niño and La Niña events in the tropical Pacific. The military is always concerned about its ability to respond rapidly to conflict situations worldwide. Thus, major powers have acknowledged that weather and climate can enhance or impede military operations. Factors related to climate and weather variability and extremes have been known for a long time to be of strategic and tactical military and defense importance. In their economic geography book, Klimm and colleagues (1940), in a section on The Value of Dependable Weather Forecasts, asked students to consider the following question: "In September 1939, at the outbreak of war [World War II], most European governments ceased to exchange weather information. Why?" It's a good question to ponder.

What next: weather wars? What about the role of weather and climate anomalies in future military operations? Wilson (1997) wrote that "the Pentagon's top meteorologists believe that the United States will be ready to fight—and win—a weather war early in the next [twenty-first] century." The unclassified optimistic Pentagon study, subtitled "Owning the weather in 2025," was based on expected advances in both weather prediction and weather modification. Apparently, some military establishments, in this case the U.S. Air Force, hope to devise ways to control or at least increase the odds of influencing atmospheric conditions from battlefield to region-wide spatial scales. Perhaps scientists will get better at techniques that tamper with atmospheric processes on time and space scales of interest to many of the climate- and weather-related sectors of society, including the military. Right now, as far as military operations are concerned, controlling the weather in combat situations remains in the realm of science fiction.

Climate Policy and Law

It is not difficult for anyone, even those who do not track news on a daily basis, to realize that when it comes to making policy and law, there are many special interests locked in heavy competition to have their views prevail in a final policy document or process. People come to the negotiating table on climate-related issues with a wide range of backgrounds, allegiances, interests, and beliefs. Cooperation and conflict often ensue in the bargaining process, and the final outputs may not be totally in any single group's interest. Global warming is one of the many issues that serves as an example of the difficulties encountered in negotiating a Law of the Atmosphere.

Global warming, of course, is the climate issue receiving the most attention today from governmental and nongovernmental organizations alike. Shogren and Toman (2000), in their paper on climate change policy, commented on the emergence of the global warming issue: "Having risen from relative obscurity as few as ten years ago, climate change now looms large among environmental policy issues. Its scope is global; the potential environmental and economic impacts are ubiquitous; the potential response—a significant shift away from using fossil fuels as the primary energy source in the modern economy—is daunting."

The international conference, Our Changing Atmosphere,

held in Toronto, Canada in June 1988 (WMO, 1988), was sparked by concern about the steadily increasing emissions of carbon dioxide and other greenhouse gases into the atmosphere and their influence on the global climate system. A 20 percent reduction in CO_2 emissions was proposed at this conference. In October 1988, the government of Malta proposed to the United Nations General Assembly a resolution on the "protection of global climate for present and future generations of mankind." In essence, this was a call for a Law of the Atmosphere. (Interestingly, the government of Malta had proposed a few decades earlier the process to develop a Law of the Sea.)

In 1988, the IPCC was jointly established by the WMO and the United Nations Environment Programme (UNEP) to assess the science, impacts, and policy implications of a greenhouse gas-induced (primarily carbon dioxide) global warming of the earth's atmosphere. It was also to provide, on request, scientific/technical/socioeconomic advice to the Conference of Parties (COPs) to the UN Framework Convention on Climate Change (UNFCCC). Following the release of the first IPCC assessment in 1990, the UN created the International Negotiating Committee to negotiate the construction of a Framework Convention on Climate Change in time for governments to consider it at the Earth Summit in June 1992 in Rio de Janeiro. (The UN Conference on Environment and Development, known as either UNCED or the Earth Summit, will hereafter be referred to as the Earth Summit.)

Since the development of the UNFCCC, national representatives of the countries that signed, ratified, or acceded to the convention have held several meetings. These meetings, referred to as the Conference of Parties, dealt with political, economic, technological, and methodological issues related to global warming. For the most part, these meetings centered on how governments might at least stabilize or, even better, reduce their nation's emissions of greenhouses gases, especially from fossil fuel use.

Since its formation, the IPCC has produced several technical and special reports and three major high-visibility assessments (1990, 1996, 2001). Keeping to its five-year timetable, a fourth assessment is planned for 2007. These assessments are intended to provide policy makers worldwide with an update on the current state of the science of global warming. Each of these reports has affirmed with increasing certainty that greenhouse gas emissions to the atmosphere are on the increase and that the global climate warmed by at least 0.5°C (0.9°F) during the twentieth century. The 1996 report was the first time that the IPCC presented the view in print that greenhouse gas emissions resulting from human activities were having a discernable influence on the otherwise naturally occurring greenhouse effect. On this point, the authors of chapter 8 in this IPCC report concluded that:

> Viewed as a whole, these results indicate that the observed trend in global mean temperature over the past 100 years is unlikely to be entirely natural in origin. More importantly, there is evidence of an emerging pattern of climate response to forcings by greenhouse gases and sulfate aerosols in the observed climate record. This evidence comes from the geographical, seasonal and vertical patterns of temperature change. Taken together, these results point towards a human influence on global climate. (Santer et al., 1996, p. 100)

This view was strengthened in the report of the 2001 IPCC assessment.

Politically divergent views on the global warming issue have led to the formation of opposing policy camps. Some issues relate to the question of who pays to reduce these emissions, whether from fossil fuel burning, industrial burning, fertilizers, or changes in land use, such as tropical deforestation. Developing-country representatives and many nongovernmental environmental groups argue that the industrialized countries are the ones that saturated the atmosphere with these gases in their drive toward industrialization. They caused the problem and, therefore, they should take the first steps to fix it. This approach is often called the "polluter pays" principle. Industrialized-country representatives and many corporate executives note that developing countries are less efficient in fuel burning and that they will become the producers of the lion's share of GHGs early in this century. Oil-producing countries, coal industries, and consumers oppose decisions that might restrict the sale of certain types of fuel or that would limit fossil fuel consumption. China has large coal reserves that it can use to pursue its economic development strategies in the absence of direct access to cleaner energy sources. Are industrialized countries ready to *give* (not sell) clean technologies to China? For President Bush and the U.S. Senate, for example, one of the touchiest issues is the desire by many countries to allow some major developing countries to increase their GHG emissions, while the major CO_2-producing countries would have to reduce their emissions.

The Conference of Parties

By the end of 2001, there had been eight meetings of the UN FCCC Conference of Parties to develop an international legal instrument to slow, if not arrest or reverse, greenhouse gas emissions.

The *first COP* was held in Berlin, Germany (28 March–7 April 1995). The delegates issued the Berlin Mandate, which called for new talks to develop a protocol to be adopted a few years later at a COP3 to be held in Kyoto, Japan. A major politically divisive issue was raised in Berlin when an Indian minister highlighted the important distinction between the "luxury emissions" of industrialized countries and the "survival emissions" of developing countries. A second challenging issue centered on a call from a delegate from China for a North-South technology transfer from the "have" countries to the "have-not" countries to produce cleaner energy for the purpose of GHG and poverty reduction. (See Eco, 2002).

COP2 was held in Geneva, Switzerland (8–19 July 1996). Its delegates were charged with preparing a protocol (to be called the Kyoto Protocol) to be negotiated at COP3. Its primary objective was to identify possible rates of GHG reductions for various countries.

COP3, held in Kyoto (1–10 December 1997), has become the most controversial to date of all of the Conferences of Parties (see box 4.2). It produced the elements of the Kyoto Protocol, calling on industrialized nations referred to as Annex 1 Parties to reduce their emissions of six GHGs by 5.2 percent below their 1990 emission levels by 2012. This COP incorporated into the

BOX 4.2.
Kyoto Protocol

In early December 1997, delegates from more than one hundred countries and representatives of scores of nongovernmental organizations and media reporters gathered in Kyoto, Japan for one of the most important environmental meetings of the twentieth century.

Delegates to the Third Session of the Conference of Parties (COP3) were attempting to move beyond the verbal pronouncements their governments had made at the Earth Summit in 1992. The trapping by greenhouse gases of outgoing long-wave radiation heats the atmosphere and leads to global climate, ecological, and societal change. This worries many governments and people because they and their economies have become used to their present-day energy dependencies and climate conditions. Scientists have suggested, for example, that with global warming there are likely to be more weather extremes, such as more intense droughts and floods. They have also suggested that tropical diseases will spread toward the poles and into the cooler regions as they warm.

Several national and regional plans were considered at COP3. The strongest one came from the European Union, which proposed a 15% reduction in CO_2 emissions below the 1990 level by 2010. It also sought to exempt developing countries from having to make any cutbacks in the near future. Much of the developing world supported the European position. However, the Republican-dominated U.S. Congress during the Clinton presidency passed a resolution stating that it would not support any Kyoto agreement that did not require developing countries to reduce their greenhouse gas emissions.

Not unlike in the industrialized world, countries in the developing world are split on the climate change issue. Small island nations fear the sea level rise that would result from global warming. This could submerge countries such as the Maldives or Tuvalu. Representatives of these nations sought immediate action to limit GHG emissions. Other developing countries appeared to be less concerned about global warming science or the proposed global warming impacts.

For the Kyoto Protocol to take effect, it must be ratified by at least 55 countries whose GHGs add up to 55 percent of global emissions in 1990.

Because this was a meeting of governments, the delegates had to produce something, even if that something was a weak agreement that would hopefully be strengthened at a later date. The Kyoto Protocol has been challenged for a variety of reasons by all points on the political spectrum. However, many countries have chosen to go forward with the protocol despite the United States' rejection of it.

Surprisingly for such an important event, worldwide media coverage preceding the conference was relatively sparse. Perhaps stories about the biggest El Niño of the century had kept the conference off the front page and the evening newscast.

A visitor to Kyoto in mid-November 1997, two weeks before the negotiations were to start, would have found little attention being paid to COP3, even by the Kyoto media. However, after the conference, increasing amounts of newspaper space around the world were devoted to the outcomes and the issues raised. Some writers in Japan called for 20 nuclear facilities to be constructed in the not-too-distant future to cover the energy shortfall that would be generated in the country by the Kyoto-proposed mandated cutbacks in fossil fuel use.

protocol the following mechanisms: tradable permits, Joint Implementation, and the Clean Development Mechanism. Industrialized countries could gain GHG emission credits for assisting developing countries and economies in transition in lowering their emissions. In addition, governments could gain credit for protecting or developing carbon sinks (e.g., tree planting to absorb atmospheric carbon). The UNFCCC press release from the conference noted that "the Protocol encourages governments to pursue emissions reductions by improving energy efficiency, reforming the energy and transportation sectors, protecting forests and other carbon 'sinks,' promoting renewable forms of energy, phasing out inappropriate fiscal measures and market imperfections, and limiting methane emissions from waste management and energy systems" (UNFCCC, 1997).

COP4, held in Buenos Aires, Argentina (2–13 November

1998), produced a Plan of Action "to reduce the risk of climate change." The conference secretariat noted that the plan "establishes deadlines for finalizing the outstanding details of the Kyoto Protocol so that agreement will be fully operational when it enters into force sometime after the year 2000." An official press release also noted that the Plan of Action addressed issues related to technology transfer and the special needs and concerns of countries likely to be affected by global warming and by "the economic implications of response measures" (UNFCCC, 1998).

The agenda for *COP5* (Bonn, Germany; 25 October– 5 November 1999) was based on the COP4 Plan of Action. The headline of the UNEP press release on this meeting captured the hopes of many of the delegates representing 166 governments: "Ministers pledge to finalize climate agreement by November 2000" (i.e., at COP6 in The Hague). The release noted that "in a further display of support for rapid action, many countries echoed the call by German Chancellor Gerhard Schroeder in his opening address for governments to ratify the protocol quickly so that it can enter into force by 2002—ten years after the Climate Change Convention was signed at the Rio Earth Summit" (UNFCCC, 1999).

COP6 was held in November 2000 in The Hague, Netherlands. More so than before, political and economic controversies surrounding the Kyoto Protocol dominated the negotiations. The U.S. position was to seek credit for its forests as carbon sinks, thereby greatly reducing the pressure on its politicians to reduce GHG emissions by 5 percent below the 1990 level. (Note: This would be equivalent to 10 percent of the 2000 level of emissions because emissions increased throughout the 1990s.) To governments and NGOs in favor of an agreement, the meeting at The Hague was a great disappointment. The election in the United States in early November 2000 of a more conservative president put a damper on progress. A UN press release captured the frustration in a veiled way: "Climate change talks suspended. Negotiations to resume during 2001."

Klaus Toepfer, executive director of UNEP, provided his diplomatic spin to the situation, stating that "it is better to suspend the talks and resume later to ensure that we find the right path forward rather than take a hasty step that moves us in the wrong direction" (UNFCCC, 2000). A news story from the British Broadcasting Corporation with the header "Climate Talks End in Failure," captured the best diplomatic doublespeak about COP6 when it quoted the Danish minister for the environment about the meeting's progress: "I wouldn't say it's a failure. It's a non-success" (*BBC News Online*, 2000).

COP61/2 convened in Bonn in July 2001 to continue the deliberations that stalled at The Hague. The U.S. government refused to take an active part, announcing that it would present a plan to cope with GHG emissions at a later date. In spite of the U.S. opposition to the Kyoto Protocol, COP 61/2 discussed institutional and financial arrangements for its implementation. The U.S. withdrawal from negotiations and the government's comments about the protocol sparked sharp criticism of its "ostrich-like" position from around the globe, and especially from European allies (figure 4.10). The proposals developed at COP61/2 were to be presented at COP7 (figure 4.11).

COP7 convened in Morocco for two weeks beginning in late October 2001. According to the FCCC's executive secretary, Michael Cutajar, "After several years of tough negotiation, the institutions and detailed procedures of the Kyoto Protocol are

Figure 4.10. Newspaper headlines from various countries attacking the Bush administration policy to abandon the Kyoto process. (ESIG, 2001)

Figure 4.11. Advertisement for COP7 in Morocco.

now in place. The next step is to test their effectiveness in overseeing the five percent cut in greenhouse gas emissions by developed countries over the next decade" (UNFCCC, 2001).

We now know that the Bush administration was serious when President Bush announced publicly in March 2001, a couple of months after having been sworn in as president, that the United States would no longer support Kyoto Protocol activities. He believed that the protocol was unworkable and unjustifiably favored major developing countries such as China, Brazil, and India.

The collection of statements taken from the official UN press releases following each of the COP meetings presents official and sanitized views on the progress and outputs of the COPs. They do not in any way explicitly reflect the specific conflicts that occurred during the negotiations. Nor do they make mention of the power vacuum that was created when the United States rejected the protocol, a political vacuum subsequently filled by the so-called "Gang of Four"—Japan, Australia, Canada, and Russia. They, each for different reasons, were able to extract concessions from other countries at COP7 that favored their own national interests (Athanasiou and Baer, 2001).

COP8 was held in October 2002 in New Delhi, India. It is apparent from media coverage of the event that little in the way of meaningful progress had been made. According to one writer, the conference was "more notable for what it didn't accomplish than what it did" (Haimson, 2002). Haimson went on to note that "developing nations successfully shrugged off an attempt by the European Union to make them commit to their own specific emission targets; the unwillingness of the developing nations to act was undoubtedly reinforced by the refusal of the United States to ratify the Kyoto Protocol on climate change."

The conference issued the "Delhi Ministerial Declaration on Climate Change and Sustainable Development," which basically highlighted the importance of adaptive mechanisms in response to climate change, as opposed to mitigative or preventive actions. The U.S. delegation acknowledged the close relationship between climate change and sustainable development.

COP9 is scheduled to take place in Italy in December 2003.

Winners, Losers, and Global Warming

Today there are climate-related winners and losers around the globe in terms of the distribution in time and space of precipitation and temperature. Some countries and areas have been endowed with favorable climate and soil conditions that support their economic development prospects. Others have not. We know about them, even if we are reluctant to name them. Yet there is little apparent, let alone sustained, effort by the climate "haves" to help the climate "have-nots" to manage their "problem climates" or their climate problems.

As the global warming issue gained greater attention in the late 1980s among policy makers, the media, the public, and special interest groups worldwide, there was a growing number of implicit as well as explicit references to the issue of those who might gain or lose from a changing climate. The comments about winning and losing have varied widely:

> Although the available information shows that moisture conditions in a number of regions have already

deteriorated due to global warming, it is probable that beginning with the first quarter of the 21st century the moisture conditions will improve everywhere. This casts doubt on the expediency of carrying out very expensive actions aimed at retarding or terminating global warming during the nearest decade. (Budyko, 1988)

Those who have been involved in this debate for a long while have come to see the winners and losers phrase as an impediment to increased awareness and a bar to meaningful action. And, in fact, I think one can make a very strong case that there are no winners at all. (Senator Al Gore, as quoted in Schneider, 1989)

Because the warming would not be uniform over the surface of the earth, it would probably produce both winners and losers among regions and nations. (Robert M. White, President, National Academy of Engineering, 1990, personal communication)

The issue of winners and losers in a global warming scenario has received very little attention from the research community. For the most part, discussions have focused on the losers. In fact, the U.S. State Department and the U.S. EPA actively discouraged discussion of these issues in the United States as late as the early 1990s. When the issue was first raised in the late 1980s and again in the early 1990s (Glantz, 1990), key government bureaucrats at these agencies viewed it as politically incorrect to discuss potential global warming winners. The U.S. State Department under the administration of George H. W. Bush challenged the convening of a workshop in early 1990 "On Assessing the Winners and Losers in a Global Warming Context." Yet, the goal of that international workshop was to identify the need for objective measures of a win or a loss that might result from global warming, based on the belief that unexpected rainfall or temperature changes alone would not be sufficient to determine a win or a loss. For example, more rain at harvest time would likely be unwanted. Thus, the idea that 10 percent more or less rain would by itself equate to a win or a loss is not a valid assumption. It depends on the setting in which the change occurs.

The dominant concern of climate scientists in the mid-1970s was global cooling and the possibility of the return of an Ice Age. In discussions about the impacts of global cooling, science writers and government agencies showed no reluctance whatsoever to name specific countries and regions as either winners or losers. For example, Ponte (1976) suggested that "adapting to a cooler climate in the north latitudes, and to a drier climate near the equator, will require vast resources and almost unlimited energy. . . . A few countries, such as equatorial Brazil, Zaire and Indonesia, could emerge as climate superpowers" (p. 238). The Impact Team (1977) suggested that with a cooling "there would be broad belts of excess and deficit rainfall in the middle latitudes; more frequent failure of the monsoons that dominate the Indian subcontinent, south China and western Africa; shorter growing seasons for Canada, northern Russia and north China. Europe could expect to be cooler and wetter. Of the main grain-growing regions, only the U.S. and Argentina would escape adverse effects" (p. 216).

The major difference between the two situations seems clear: a global cooling would be a naturally occurring phenomenon and, therefore, no particular government or country could be blamed for its onset or its adverse impacts. For global warming, however, industrialized countries that benefited from the Industrial Revolution (say, from 1760 to 1830) and beyond are primarily responsible for having saturated the global atmosphere with a variety of greenhouse gases. Thus, one could argue that they bear the responsibility for a human-induced enhancement of the greenhouse effect and its potentially adverse consequences, which would most likely harm the developing countries to a greater extent. Hence, an attitude among some government officials and corporations in industrialized countries has been to portray everyone as a loser in a global warming scenario. This is only true with regard to sea level rise; for this environmental change there can be no winners, unlike for shifts in rainfall, for example.

More than ten years have passed since the workshop on winners and losers was held, and there is still need to develop objective measures of what it might mean to be a climate winner or climate loser, whether or not the global climate changes in a profound way. The need for such measures is greater now than ever (figure 4.12). Some countries are engaging in activities that yield short-

Willy 'N Ethel

Figure 4.12. Cartoon reprinted with permission from North American Syndicate.

term benefits, which will later cause much greater costs. The United States' rejection of the Kyoto Protocol is one well-known example. Another recent example relates to Russia.

The Russian Federation is keen to exploit its natural gas fields in its polar regions. As the sea ice melts in the Arctic, it becomes easier for the gas industry to explore and exploit the tapping of its gas resources. That is the good news, or the "win" part of the equation. The bad news is that the increased use of natural gas along with other fossil fuels is enhancing the human-induced global warming of the earth's atmosphere. This, some believe, has led to droughts and forest fires throughout Russia. Thus, the feedback is a deadly one for the Russian economy at present, and especially in the long term (Drillbits & Tailings, 2000).

Getting to Yes on Ozone Depletion: The Montreal Protocol
Chlorofluorocarbons are human-made chemical compounds developed in the 1920s for use as refrigerants, and later as foam-

blowing and propellant agents and as cleansing agents in the electronics industry. In the mid-1970s, atmospheric chemists discovered that these chemical compounds, although inert in the lower atmosphere, are highly destructive of ozone once they make their way into the upper atmosphere layer known as the stratosphere (Molina and Rowland, 1974; Stolarski and Cicerone, 1974). There, the molecules break down because of the ultraviolet radiation, thereby freeing the chlorine atoms to interact with and break down ozone molecules. The free chlorine atom becomes an "ozone eater," with each atom capable of destroying thousands of ozone molecules, which in the stratosphere protect life on earth from lethal doses of ultraviolet-B radiation (UV-B). CFCs are also considered a greenhouse gas, thereby contributing to global warming.

Although some governments took immediate action in the late 1970s to limit CFC use as propellants in spray cans, others contended that there was no solid scientific proof about the destructive nature of the CFCs. Companies that relied on using CFCs, with very few exceptions, fought legislation to curtail CFC production, complaining, for example, that no good substitutes existed for CFCs as refrigerants or for its other specific uses. The British chemical industry commissioned an atmospheric scientist to tour the United States in the late 1970s to challenge the proposed bans on CFCs. These chemicals became a domestic and international political issue.

When British scientist Joe Farman discovered a sharp thinning of stratospheric ozone over the Antarctic (since referred to as the "ozone hole"), political action to reduce CFC manufacture, use, and trading was reinforced if not accelerated. In March 1985, governments met in Vienna, Austria to develop a framework convention for the protection of the ozone layer. Chasek (2001) wrote that this convention "was essentially an agreement to cooperate on monitoring, research, and data exchanges. It imposed no specific obligations on the signatories to reduce production of ozone-depleting chemicals and did not even specify what chemicals were the cause of ozone depletion" (p. 104).

Additional intergovernmental meetings were held, and new scientific evidence was presented over the next couple of years, setting the stage for negotiation and acceptance of the *Montreal Protocol on Substances that Deplete the Ozone Layer* in 1987 (and later amended) (see Benedick, 1991). Perhaps one of the most important aspects of this particular political issue was that it was addressed in the absence of statistical evidence of impacts on human health. It was in essence an application of the precautionary principle (see "Precautionary Principle" in Climate Ethics and Equity section).

The Montreal Protocol has led to the sharp reduction of the worldwide manufacture and use of CFCs. Given that CFCs have an estimated residence time in the atmosphere of a century or more, the Antarctic ozone hole remains; another ozone thinning has appeared in the Arctic region.

An illegal trade has developed in CFCs, because several countries have agreed to ban production or use only in 2005. In 1996, Russia, China, India, several in Eastern Europe, and a few other countries were identified as sources of CFC smuggling activities (See Witte, 1996).

Climate-Related Disaster Diplomacy

Ever since the Turkish invasion of Cyprus in July 1974 and the division of that island into two political parts, Turkey and Greece (both NATO allies) have been at political and military odds. However, in 1999 an apparent shift took place in the decades-old hostile relationship between them. Turkey suffered from a devastating earthquake; tens of thousands were killed and hundreds of thousands displaced. Greece, in a gesture of goodwill, offered humanitarian assistance to its enemy. The Turkish government was appreciative (Hope, 2001). In a twist of fate, a few months later Greece suffered as well from the impacts of an earthquake, although not as devastating as the one in Turkey. Turkey offered assistance to Greece. Since then, there appears to have been a slight warming of relations between these countries. Could one refer to their interactions, inspired by earthquakes, as "disaster diplomacy"?

However, in most circumstances, disasters do not have a positive effect on the international relations between enemies. For example, one might think that countries such as the United States and Cuba, which are destined each year to face the possible adverse impacts of a shared natural hazard—the hurricane—would be prompted to work very closely together to forecast, monitor, and respond to hurricanes and their underlying causes. The United States and Cuba might also work closely together to better understand La Niña events because there are more hurricanes in the tropical Atlantic during a La Niña event.

In fact, U.S. and Cuban hurricane specialists do work closely together when a specific hurricane threatens their countries. They also interact to some extent in training forecasters. When it comes to hurricanes as a research topic, there is less cooperation. As for research on La Niña/El Niño hurricane linkages or on global warming issues, interactions usually take place either by accident or informally at international conferences, rather than on a planned bilateral basis. This lack of cooperation is because of the animosity between the Castro regime and successive American governments. Ill feelings remain so high that they overshadow any potential goodwill generated by a shared disaster that might lead to a warming of diplomatic relations (Glantz, 2000).

In 2000, severe drought conditions prompted the Cuban government to request food assistance from the international community for the first time since Castro took power in 1959. The U.S. government said it would consider Cuba's appeal, but Cuban authorities declined any assistance from the United States, stating that there would be no food shortages if it were not for the United States' multidecade trade embargo. In that same year, the U.S. government modified its laws with respect to trade with Cuba, but purchases by Cuba had to be paid with cash (Pianin, 2000).

In early November 2001, Hurricane Michelle inflicted considerable damage on Cuban food production systems and, once again, Cuba sought food aid from the international community. This time, the United States offered to sell food, and the Cuban government opted to buy American corn and soybeans, noting that it was going to be only a one-time event (DePalma, 2002).

Climate Economics

The financial aspects of climate are quite clear for those interested in vacations or for those who operate climate-sensitive vacation resorts. People do not plan skiing vacations when there is little chance of snow at their destination or plan a beach holiday in the sun where the sun is weak or only infrequently visible. The same is true for climate-related hazards such as droughts, floods, fires, or frosts: people think mostly about the financial impacts, such as lost crop sales, damage to the built environment, costs of rebuilding, lost business, insurance payouts, and so forth. Climate economics really encompasses much more than finance, however. Mainstream economics relates to the production, distribution, and consumption of goods, services, and environmental amenities. Climate economics relates to how one's well-being might be influenced directly and indirectly by climate variations on all time scales.

Physical scientists seem to have an affinity to working with economists, as opposed to other social scientists. Economists are quantitative and can provide them with estimates of the costs and benefits of extreme events or the value of forecasts. They could help policy makers to identify winners and losers in various global warming scenarios. Besides, like scientists, economists use models and develop scenarios based on various assumptions and indicators.

When people think of climate economics, they most likely think about the cost or benefit of a change in the quantity or quality of economic goods and services associated with climate or climate-related impacts. To do this properly, however, researchers, regardless of academic training, have to develop a better understanding of "attribution." This is the process of identifying what impacts can appropriately be blamed on or associated with a climate anomaly or a weather extreme. A climate anomaly never affects a society in isolation. The society may simultaneously be affected by other factors, such as poverty, inappropriate land use, conflict, transboundary air or water disputes, rivalries among bureaucratic units, domestic political rivalries, and so forth. A researcher must take great care to separate out which impacts can be attributed to a climate anomaly and which are due to societal activities. Sometimes a climate anomaly by itself might not have been the most important factor but would become so when combined with other physical or social factors.

Economic assessments are of major interest to societies and governments on time scales ranging from seasonal to long-term. Costs and benefits always accrue to someone, regardless of the climate phenomenon of concern. For example, when a weak El Niño is forecasted, people in North America might feel less threatened. However, a weak El Niño can still have major consequences in countries on both sides of the equatorial Pacific, such as Australia and Peru.

Farmers have to cope each season with considerable uncertainty surrounding their agricultural activities: Will they be able to plant on time in the spring? Will it rain when it is time to apply fertilizer and pesticides? Will there be enough rain in the summer? Will it be dry at harvest time? Will the weather be so favorable for agriculture that a bumper crop results, causing a drop in earnings for the farmer? The truth of the matter is that farmers do not want perfect weather for their agricultural activities. Perfect weather would mean high yields, small losses in harvesting, large production surpluses, and low prices in the marketplace. Perfect weather could be and has been a financial disaster to individual farmers and an economic hazard for many governments, which earn much of their foreign exchange from the export of agricultural products. Micro- and macroeconomic impacts of weather and climate have been studied for decades and are in a general sense well known.

At the micro-level, farmers are in competition with other farmers growing the same crops. Each farmer hedges his or her bets on which crops will survive a growing season of unknown characteristics and provide an income sufficient to support the family's needs. Often, farmers diversify their crop portfolios and have developed their own set of forecast indicators that they consider reliable. They respond to those indicators throughout the growing season but are heavily influenced by forecasts of official government agencies.

Herders in the African Sahel were once accustomed to moving their livestock with the progression of the seasonal rains. They are forced to move from one shallow well or rain-filled depression to another, as water in the wells and depressions dries up. About the same time that happens, the vegetative cover is well grazed and the soils trampled around the water source. With the widespread use of deep-tube well technology during periods of drought in sub-Saharan Africa, permanent human-made water resources were created artificially in arid areas. Herders, lulled into a false sense of security, would keep their livestock near the deep wells. In time, however, overgrazing would reduce the availability of vegetation. Then, the diminished vegetation would be out of balance with the now-permanent water supply. Because the rains had long since passed in the area, little vegetation was left for the livestock to eat before they could migrate to the next well. Hence, the animals would perish in great numbers as, for example, happened during the 1968–1973 drought in the West African Sahel, as often as a result of hunger as of thirst (Glantz, 1976).

Starting in the mid-1970s, the perceptions of economic development specialists shifted from viewing climate as a boundary constraint that limited a society's ability to prosper to viewing climate as a variable environmental condition with which a society could cope, given proper institutional arrangements. Today, many governments base their future expenditures for economic development activities on the expected earnings from the export sales of their weather- and climate-sensitive products. Governments whose economies are dependent on the export sales of climate-sensitive products, such as agricultural products or fish, are also influenced in a major way by trade policy constraints, some of which relate to climatic factors, imposed by potential importers of those products.

Everyone affected by climate is living with uncertainty in his or her activities. Uncertainty surrounds weather and climate anomalies. As a result individuals, corporations, and governments have increasingly learned of the need to build hedging strategies and tactics into their activities to better cope with those anomalies.

Externalities

"Externalities" refer to the costs of environmental degradation that are seldom taken into account in traditional cost-benefit analyses. For example, the adverse impacts of downwind or downstream environmental pollution are often not taken into consideration as a cost of a particular industrial or agricultural activity. This omission tilts the cost-benefit assessment toward a favorable benefit and away from the more realistic costs. Meanwhile, those who live downwind or downstream must bear the environmental and health costs of that pollution. As ecologist

Barry Commoner once noted, when it comes to pollution, "there is no free lunch"; someone, somewhere, and at some time has to pay for that pollution. Often, it is the victim who ends up paying.

As a result, policy makers developed the polluter pays principle, which proposes that polluters are responsible for the adverse impacts of that pollution wherever and whenever they occur. Although such a notion makes sense to the person on the street, the lack of scientific certainties that can link cause and effect of pollution damage often block successful legal action against the polluter. Unfortunately, polluters often hide behind scientific uncertainties to avoid penalties. Several Hollywood feature films have been based on the legal battles between polluters and the polluted (e.g., *Erin Brockovich, A Civil Action, Karen Smallwood, Toxic Avenger,* and *Fire Down Below,* to name a few).

Risk takers and risk avoiders exist in all societies, whether those societies are rich or poor, industrial or agrarian, totalitarian or democratic. Risk takers are willing to take chances based on the information at hand, even if it is uncertain or incomplete. They are often aware, at least in a general way, of the probabilities associated with the risks that they take. Those who are risk-averse, however, are more conservative about changing the status quo. Risk makers are an as yet unrecognized category. Risk makers create negative externalities for the rest of society, the costs of which society must bear in the form of insurance subsidies for coastal development, costs of greenhouse gas mitigation, and so on. The consequences of decisions made by risk makers do not affect them directly. Their decisions create risky situations for others, who then have to bear the consequences. An example of risk makers would be decision makers who allow people to develop settlements in floodplains or on unstable mountain slopes. In such situations, policy makers knowingly allow, if not encourage, others to move into harm's way. Lack of awareness about the range of climate sensitivities in a specific geographic region or socioeconomic sector can be considered a form of risk making.

Discounting

Discount rates have two basic meanings in economics. The first relates to the cost of an activity based on the current value of currency, that is, the opportunity cost of capital. Economists set an annual rate of decline in value to be about equal to the contemporary interest rate on capital. Discounting is often used to calculate the economic value of using a resource now as opposed to keeping it intact for use by future generations. Its future value is always less than the present value.

A second basic meaning of discount rates depends on preferences and is subjective. It is based on the belief that a dollar today will be increasingly less valuable as time progresses, but the rate of depreciation of that dollar will depend on a person's time preferences. Discounting is "a method used by economists to determine the dollar value today of costs and benefits in the future. Future money values are weighted by a value of less than 1, or 'discounted.' This reflects a commonly observed preference of individuals for reaping benefits sooner rather than later while delaying costs" (Toman, 2001, p. 267).

People, economists included, also discount the past, that is, they do not put as much value on information derived in the distant past as in the recent past. For example, experiences of our fathers, let alone our grandfathers, are no longer considered as relevant as they once might have been. Although the recent past may have some influence on our climate-related decisions, the distant past does not. Discounting the past is based on the subjective view that we are more advanced today than before and that new knowledge, new technologies, and new coping strategies have made past experiences less relevant. This is in direct conflict with the adage that "those who do not learn from history are doomed to repeat it." Jamieson (1988) has suggested that using experiences in dealing with climate impacts in the recent past can provide societies with a glimpse of their futures, in the absence of reassessments and rethinking about those experiences. Despite the existing wealth of economic, political, and cultural lessons gained in the past about climate-society-environment interactions, many of those lessons remain unused and unknown to the present generation of decision makers.

Intergenerational Issues

Whose welfare should receive a higher value from today's decision makers, present generations or future ones? What sacrifices are present generations willing to make for future ones or for past ones—for example, by protecting cultural capital such as heritage sites? These are some of the major issues in the global warming and tropical deforestation debates. Two opposing mindsets exist: (1) what can I do so that future generations can enjoy the global climate regime of today (rather than handing the future generations a hotter climate and its associated extremes); and (2) why should I do anything for future generations? What have they done for me? One might also argue that he or she has no idea what future generations might prefer. How one views his or her obligations to future generations will clearly affect resource use today. Our generation did not start the global warming problem; it inherited it from previous generations. Should we deal with it now and sacrifice, even if the results of our efforts might not be visible until many decades in the future? Can we compensate future generations for any harm we might inflict now through, for example, enhanced economic growth? These questions are of major interest to environmental ethicists as well as to economists and should be of concern to all policy makers.

Climate Variability and Economic Development

Today, societies are attuned to the natural flow of the seasons in their regions. Any disruption to that flow, as a result of war, drought, pestilence or flooding, can cause great delays in progress toward economic development. As noted earlier, a drought, for example, can take place in one growing season, but its impacts on society could linger for years. Although many societies are able to get through a year of drought, they have much more difficulty getting through two or three years of consecutive drought, and multiyear droughts can severely damage an economy. Droughts have led to the abandonment of several settlements. Floods, too, have had similar long-term consequences for societies.

Another example of the influence of variability on development activities occurred in the early 1950s, when the California sardine fishery finally collapsed after showing signs of fragility under fishing pressure and decline over several decades. The controversy continues today as to whether the demise of this lucrative industry was the result of overfishing or the result of a climate-related fluctuation every few decades in the marine environment. In any event, the raw material for California's fishery, the sardine, disappeared in the early 1950s.

Interestingly, several of the idle boats and fish-processing factories were exported to other countries in the mid-1950s, primarily South Africa and Peru. Until then, these two countries were not important fishing nations. However, the influx of cheap fishing-industry infrastructure opened the living marine resources in their coastal waters to all-out exploitation. South Africa developed a multispecies fishery. Peru embarked on the capture of anchovy, using it as an industrial (as opposed to a food) fish. Fishmeal production became a growth industry, as catch doubled each year during the 1950s. The demand grew markedly for fishmeal as an animal feed supplement in the United States, for the poultry industry mainly, as increasingly affluent Americans shifted their diets from grains to meats. By the late 1960s, Peru's economy derived about 30 percent of its foreign exchange from the export of anchovy fishmeal. A combination of factors, not the least important of which were the 1972–1973 El Niño and overfishing, brought about the collapse of the anchovy fish population in Peruvian waters. When the fishing industry collapsed, that revenue was lost to the government for almost two decades. It picked up again only in the early 1990s. Unemployment and social unrest ensued for several years after the collapse (Glantz, 1979). The South African fisheries went through a similar boom-to-bust cycle in the 1970s and 1980s.

In December 1999, floods and mudslides in Caracas, Venezuela, killed upward of 50,000 people. This captured the concern of the Venezuelan people, politicians, and development specialists. Victims of Hurricane Mitch (1998) are still suffering from its impacts and marginal assistance from the international development community.

The multiyear drought in Afghanistan, combined with the agricultural policies of the repressive Taliban regime, led to near-famine conditions within the country in 2001. The impacts of this prolonged drought put considerable pressure on neighboring countries when refugees sought safe haven from food shortages as well as from political repression.

The examples above underscore the fact that climate considerations are important to managing a wide range of natural resources and the methods that societies use to exploit them. Weather or climate anomalies not only have adverse physical and societal impacts but also psychological impacts that can last for several years, if not a lifetime.

Cost-Benefit Analysis . . . of Human Life

While carrying out the assessment for the 1995 IPCC report, economists were asked to put a cost on the social value of a climate change. These numbers were apparently needed to calculate the actual costs of a global climate change based on some scenarios of the future level of warming and its potential impacts on environment and society. They used as a measure the projected impact on human mortality of a rise in average temperature around the globe of 2°C (3.6°F). The mainstream economists (among others) believed that to get a good idea of the benefit of avoiding impacts versus the cost of abatement of GHG production, an economic value of a human life had to be calculated in monetary terms, i.e., U.S. dollars. A group of economists took the challenge, made the calculations, and came up with the desired numbers as requested. They calculated the value of a human life in an industrialized country to be on the order of $1.5 million, whereas the value of a human life in a developing country was calculated to be $150,000 (figure 4.13).

These calculations generated a virtual firestorm of controversy and criticism. They raised questions not only about the monetary difference in the costing out of the value of human life but also in the ethical aspects of undertaking such a task for the sake of quantification. The issue raised serious doubts about the objectivity and representativeness of the IPCC process.

Meetings were convened to address how the calculations were derived, to question the IPCC rules given to the lead authors of various chapters in the report, and to challenge the cost/benefit approach to the issue of the social cost of climate change. Several articles appeared that attacked the authors, the numbers, the IPCC, and the ethics of having made such calculations in the first place. The analysis has been called absurd and discriminatory and was referred to by John Adams in *New Scientist* as "the economics of genocide." Despite the opposition, the numbers remained in the final report. It left a proverbial black mark on the IPCC's objectivity.

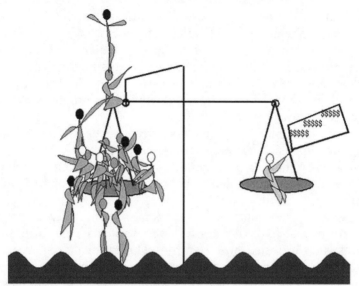

Figure 4.13. Are the lives of fifteen Bangladeshis worth the same as only one American? "This was the assumption of economists who created the global cost-benefit analysis of climate change for the IPCC. Global Commons Institute led the campaign to defend the value of life by rejecting this crazy analysis." (Source: GCI, 1996)

The Global Commons Institute in the United Kingdom was a major challenger to this social cost approach. It identified "several ways of calculating the damage costs, and thus showing that there is such a wide range of possible results that the cost-benefit approach is effectively doomed to failure" (GCI, 1996).

Climate Ethics and Equity

"Climate ethics and equity" is a phrase that was developed for the notion of climate affairs. Ethics is defined as a principle of right or good conduct or a body of such principles. It is also defined as a system of moral principles or values and as the study of the general nature of morals and specific moral choices. When applied to climate and climate-related issues, one can identify numerous ethical and equity considerations. For example, should polluters of the atmosphere pay to clean up their pollution or compensate those affected by it? Although decision makers in the United States might support this concept when applied to situations in which they are not directly involved, they appear reluctant to support the notion when they are the polluters. This is the case with regard to the inordinate share of greenhouse gas emissions produced by the United States.

The United States, with about 4 percent of the world's population, is responsible for producing about 25 percent of the global emissions. Yet, our government leaders have shown great reluctance to take responsibility as a major contributor to global warming.

As another example, is there a moral responsibility for those countries that have climates favorable to sustained agricultural production to assist those countries whose climate regimes are more problematic for sustained and reliable agricultural production? At least in theory, the international transfer of technology and information could accomplish this if the productive countries were willing to engage in such transfers.

When drought or some other climate-related disaster occurs, the impacts are not borne equally across all regions or classes in society. In fact, it is usually the poor who bear the brunt of such impacts. They live in the marginal areas. They occupy the floodplains and the unstable hillsides. The reasons behind the differential effects of climate and climate-related impacts are fair topics for those interested in climate ethics and equity.

Ethical and equity issues are not just related to the impacts of climate but also to the dissemination of climate information, including forecasts. When an organization such as the U.S. National Oceanic and Atmospheric Administration's (NOAA) Climate Prediction Center issues a forecast of the onset of an El Niño, it may be placed on the Internet in English. Who, then, can receive and use this information? As one might imagine, it would be those individuals, groups, and corporations that are able to read English and have the resources to connect to the Internet and receive NOAA's forecasts (i.e., they have a computer, are linked to the Internet, and know which sources are credible). Forecasts from various government agencies are not necessarily produced in other languages or presented in other media (radio, television, local newspapers). For example, the well-to-do fishermen in Peru will receive such forecasts well in advance of the poorer artisanal fishermen who have to rely on local newspapers or radio announcements for El Niño-related information.

Before gaining access to the Internet in the mid-1990s, researchers in northern Peru, a region subjected to torrential rains and flooding during El Niño, were totally dependent on El Niño-related information and forecasts passing through the country's scientific community in Lima, a region generally unaffected by El Niño. However, once researchers in northern Peru were able to access the Internet, they were then in a position to seek their own contacts and to analyze El Niño-related data for themselves.

Weather and climate information does not necessarily create inequities but most likely tends to enhance inequities that already exist within and between societies and sectors of society. Although such information is often used to reinforce the status quo, it can also be used to change it.

An Inter- vs. Intragenerational Equity Conflict

Governments everywhere face the same question when confronted by a climate issue: From an ethical viewpoint, who should be helped in the event of climate or climate-related problems? Many people in the present generation do not believe that they owe anything to the future. Given this perspective, there would be little incentive to preserve natural resources for use by future generations. On the other hand, many people do desire to pass something of value on to future generations, including their children and grandchildren. They seek to manage present-day natural resources while keeping in mind the needs of future generations. Examples of this intergenerational conflict can be found in all parts of the globe.

Should the Brazilian government preserve Brazil's tropical rainforests, for example, for potential discovery of pharmaceuticals, or cut them down to clear the land for cultivation and livestock and sell the timber to Japan? What is the value of this forest, today or in the future, to the earth's environmental health? Some people have called the rainforest in the Amazon the lungs of the earth. Many Brazilians are poor and in search of land for farming to grow subsistence crops or raise livestock. Researchers who deal with the notion of sustainable development are constantly grappling with the inter- versus intragenerational conflict. That is why there are so many competing views about what sustainable development means and how to achieve it—and for whose benefit. Balancing the real needs and pressures of the present generations with the potential needs and interests of future generations is an issue deserving much more attention from disciplines other than philosophy.

In reality, several generations coexist at the same time. The present generation of policy makers could ask questions and discuss climate issues (pollution, climate change, ozone depletion, deforestation, etc.) with representatives of both "past" and "future" generations (i.e., octogenarians and teenagers). Perhaps convening such an intergenerational meeting focused on climate concerns would go a long way toward understanding how environmental degradation came about and how the next generation of decision makers might cope with it in the future.

Precautionary Principle

Since the late 1980s, the notion of a precautionary principle has made considerable headway in political discussions about dealing with environmental and health issues. Because the idea has generated so much controversy over how to apply it objectively, it could in itself qualify as a policy issue. However, the principle also raises legitimate ethical and equity concerns. With respect to the environment, the principle calls for restraint on activities that could conceivably cause environmental harm. A 1998 Wingspread conference (SEHN, 1998) identified four main components of the principle, each of which has been challenged by those who oppose it:

1. Act to prevent harm despite uncertainty.
2. Shift the burden of proof to proponents of a potentially harmful activity.
3. Examine a full range of alternatives to potentially harmful activities, including taking no action.
4. Democratic decision making should be followed to ensure inclusion of those affected.

BOX 4.3.
Hurricane Mitch and Generational Issues

In late October 1998, the worst hurricane in 200 years, so experts say, made landfall in Honduras and adjacent countries in Central America. Hurricane Mitch did most of its damage in Honduras, after it had already been downgraded from an intense hurricane to a tropical depression. Unfortunately, it was a slow-moving system that dumped large amounts of rainfall along its path. Torrential rains caused flooding and mudslides, as soils became supersaturated. Villages were inundated; banana plantations and agricultural fields were flooded and crops ruined; roads, rails, bridges, and communications networks were destroyed. In Honduras alone, an estimated 10,000 people died and another 7,000 were declared missing. El Salvador, Nicaragua, and Guatemala also suffered a considerable number of deaths, damage to the natural and built environment, and social disruptions. Tens of thousands of people migrated away from the affected region, northward to the United States and southward to Costa Rica. In addition to the influx of destitute Nicaraguan refugees, Costa Rica was also directly affected by the hurricane.

In the aftermath of Mitch, the international donor community, both governmental and nongovernmental organizations, responded with emergency assistance to the victims. Financial pledges were offered and it appeared for a short while that things in the region would soon be brought back to normal. However, pre-Mitch "normal" in Honduras was not good: it was the fourth-poorest country in Latin America, corruption was widespread, the agricultural sector was in need of land reform, and the gap between the rich and poor was large. The government, business sector, and aid organizations had to decide, for example, whom to help and where to rebuild and whether to restore the banana plantations. The inherent conflict between two often disassociated foreign assistance objectives—emergency disaster response and sustainable development—was highlighted.

At the risk of over-generalizing disaster responses, the tendency to get back to normal prompts relief agencies to put up tents where the houses had been and trailers where the schools had been. However, doing so puts people back in harm's way in the event of a future extreme rainfall anomaly. Once the disaster response teams leave, having achieved their short-term goals, those responsible for long-term development assistance enter the picture. Those people found that the flexibility of their sustainable development planning had been sharply restricted by the desires of the disaster response community to get life back to normal. Thus, an ethical dilemma becomes apparent: does one help immediate survivors of Hurricane Mitch or does one focus on helping future generations avoid a similar risk decades in the future by not letting the situation "return to normal?" This is a conflict between future generations and those alive today. Is there a way to accommodate the needs of both groups?

Commercial fishing activities provide a good example. Many commercially exploited fish stocks have collapsed under a combination of fishing pressure and natural factors. The question often asked is whether the stock might have survived had it not been overexploited. Fishery scientists have, over the past several decades, suggested fishing guidelines to exploit fish populations: maximum sustainable yield (MSY), optimal yield, and safe yield. The last one calls for a light exploitation of a variable fish population to avoid a collapse of the fish stock. At the other extreme, MSY perpetuates a high level of exploitation, putting the fish stock at a great risk of collapse, given the scientific uncertainty surrounding fish population dynamics and the marine environments.

The precautionary principle is like the adage, "look before you leap." It makes total sense to those who seek to prevent (if not roll back) activities that can be destructive to the environment. The basic tenet of the precautionary principle is "do no significant harm." Who could challenge the proposition of putting safety considerations first? Nevertheless, there is considerable opposition to placing an unquestioned reliance on the precautionary principle as a litmus test for society's interactions with the environment. One writer (Bailey, 1999) has noted, "For some people in some situations, 'look before you leap' is good advice. Others might be wiser to heed the equally proverbial 'he who hesitates is lost'" (p. 8).

The principle has many detractors and many supporters. For example, governments that don't want to take action (e.g., President Bush on the global warming issue) can delay their need to do so by funding more research in the guise of reducing scientific uncertainty. Corporations, too, that depend on the exploitation of the natural environment or engage in industrial processes that adversely impact the earth's environment, such as the chemical industry, also tend to call for more research, playing up the scientific uncertainties surrounding environmental changes resulting from human activities. This is exactly what happened regarding the spray can-CFC-ozone depletion debate of the late 1970s.

The problem and controversy arise over the fact that invoking the principle calls for stopping potentially harmful activities before scientific investigations about cause and effect have been completed. So decisions that invoke the precautionary principle are dealing with scientific uncertainty about impacts on the environment. Clearly, the principle is not value-neutral, nor is opposition to it.

Review of the pros and cons of the precautionary principle applied to climate-related issues can expose important strengths and weaknesses in generally accepted principles and concepts used to make decisions about environmental change (e.g., risk assessment and cost-benefit analysis).

A recent report from the European Environment Agency (Harremoës et al., 2002) highlighted the value to policy makers of using this principle in their decisionmaking processes. Entitled "Early warnings, late lessons," it presented more than a dozen case studies that showed how scientific uncertainty had been successfully used over decades to block policy changes related to the use of chemicals that over time proved to be very harmful to the environment and people.

The precautionary principle has grown in popularity since the late 1980s. Despite political opposition to it, the principle is increasingly being taken into account in many environmental agreements and is now mentioned with regard to climate change issues. It calls for active mitigation in the face of scientific uncertainty about the possible effects of human-induced climate change.

North-South Divide and Climate Change

International discussions about global warming have already identified ethical and equity aspects of the climate change issue, many of which reflect North-South socioeconomic, cultural, and political differences. The North-South Divide refers to the economic development gap that exists between the industrialized countries of the mid- and high latitudes and the developing countries, primarily in the tropics and subtropics. For example, developing countries argue effectively that the industrialized countries developed their economies and societies as a result of their dependence on the burning of fossil fuels in the process of industrialization. Thus, they argue that the industrialized countries have an obvious responsibility to take the first step to unconditionally reduce greenhouse gas emissions. The industrialized countries, in turn, argue that the developing ones have a responsibility to limit their growing greenhouse gas emissions, now that the global scientific community has linked such emissions to serious changes in the global climate system. They argue that the developing countries will be emitting the lion's share of the greenhouse gases in the near future. Developing-country representatives believe it is unfair to ask their countries to forego economic development because of a problem initiated centuries earlier by developed countries.

Then there is the problem of sea level rise and the inundation of small island states in the Pacific. Is there an obligation on the part of the major greenhouse gas-emitting nations to do all in their power to avoid such an eventuality? Examining per capita GHG emissions raises equity issues because it is well known that people in developing countries cause far less carbon dioxide per capita to be emitted than those in industrialized countries. As one can see, many ethical and equity issues are embedded in North-South positions over global warming.

Several societal sayings make ethical aspects of climate explicit. The polluter pays principle stems from the view that "if you do the crime, you should pay the fine." Often, governments and corporations apply this principle to others but are reluctant to accept responsibility for environmental harm they cause. As another example, the phrase "common but differentiated responsibilities" attempts to explicitly address the ethical and equity issues related to different levels of responsibility between industrialized and developing countries for their human contribution of greenhouse gas concentrations.

Environmental Justice

Why is it that poor people in a society are usually the most adversely affected by a drought or a flood? They are the ones whose living space has been relegated to marginal areas such as floodplains, steep hillsides, arid lands, swamp lands, and areas with precariously short growing seasons. Seldom are the upper classes in a society in harm's way to the same extent as the poor. The rich have options that the poor do not have to avoid climate-related harm. When torrential rains come, shantytowns are often washed away by rivers that overflow their banks or by mudslides. The poor have less access to medical facilities in the case of climate-related emergencies and are more frequently subjected to climate-sensitive diseases because they often lack indoor plumbing and access to clean water. Thus, there is a clear link between death and destruction resulting specifically from climate anomalies and the level of poverty in a given area.

Drawing Down Nature's Capital

Everyone is familiar with the concept of banks. If you want to borrow money, you go to a bank and request a loan. Both the banker and the borrower expect that the loan will be repaid at some time in the future. Even if it is an interest-free loan, the principle will have to be repaid. Consider the natural world a bank. Nature's bank consists of a finite amount of natural resources. Those resources can be qualitatively different: a pristine rainforest or a clear-cut one; fertile soils or soils depleted of nutrients; clean or contaminated water; and abundant fish populations or ones decimated by overfishing.

Using a bank as an analogy, one could effectively argue that the industrialized countries have drawn on nature's quantitative abundance of natural resources to attain their high levels of economic development and that they have also depleted its qualitative abundance. Because they filled the atmosphere with trace amounts of greenhouse gases from fossil fuel burning and CFCs from various industrial and domestic purposes, it is time for them to pay back nature's bank. This would require restoring the environmental quality that they "borrowed" over centuries to reach their present-day level of prosperity. This would then enable the poorer countries seeking economic development to use the natural environment's qualitative aspects without producing a net increase in the total amount of global pollution.

To take a loan from a bank without expecting to repay it would be a form of robbery. By analogy, many argue, to rob nature's bank of its quality with no intention of paying it back is also a form of theft. In this situation, they say, shouldn't the industrialized countries do the right thing and repay nature's bank the outstanding loan it took out from global environmental quality?

Clearly, ethical aspects of climate and climate-related issues are both numerous and important for societies to address to achieve equity and transparency and to entrain civil society in vulnerability reduction. A representative of an environmental NGO in Pakistan identified some of the potential contributions of a focus on climate ethics. "Climate ethics is a notion that could lead to human rights development and to eco-justice. It provides an opportunity for NGOs and others to be activists, an ethics and equity watchdog of sorts, rather than 'relativists.' It makes explicit the human feeling (emotion) of climate-environment issues. It can provide climate issues with an emotional content that is sometimes missing from physical and social science research. Climate ethics considerations could give additional breadth and depth to climate-related issues of concern to philanthropic organizations involved in social and economic development" (personal communication, Asad Abbas Naqvi, 2002).

Could Weather Affairs Be Far Behind?

Unseasonable weather and inadequate stores worried the English captains as much as they did the Spaniards. (Mattingly, 1959, p. 257)[*]

During a recent discussion with weather researchers about extreme events, I referred to "climate," and members of the audience asked me why. They said in no uncertain terms that they were interested in weather and not in climate; they were interested in atmospheric processes at time scales of up to 7–10 days, two weeks at the most. Later, at a symposium on El Niño, a well-known expert displayed his personalized pen, which was inscribed, "we do climate, not weather." Such a distinction may be acceptable to experts, but for nonspecialists climate and weather are blended together. It is not possible to talk about one without making reference to the other. Although those interested in meteorological phenomena that occur on shorter time scales and smaller space scales may not be attracted to the notion of climate affairs, they might find "weather affairs" interesting. Aspects of weather affairs could for the most part parallel those of climate affairs.

Briefly, any explanation of *weather science* must begin with a description of the global climate system. This would be followed by a description of the structures and functions of various types of weather systems: tornadoes, hurricanes, frosts, freezes, and fronts, etc. Weather variability and weather extremes are very important to most sectors of society, because most societies have attuned their socioeconomic activities to their locale's climatic characteristics, which are based on weather phenomena. It is also useful to look back through history to identify how different civilizations over millennia viewed their weather and its anomalies.

Weather variability and extremes occur on short time scales of minutes to a couple of weeks. They can have tremendous impacts on the atmosphere as well as on terrestrial, coastal, and marine ecosystems. They can also have considerable impacts, for good or

---

[*]Garrett Mattingly wrote about the Spanish Armada covering 18 April–30 July 1588.

ill, on all aspects of human settlements. For example, a torrential rainfall lasting just a few hours can wreak major havoc on a society for a long time. Hurricane Floyd hit North Carolina a few years ago, and some of the poorer communities have still not been rebuilt. Forecasting the occurrence of weather extremes has been quite challenging but when successful has yielded many benefits, such as lives saved or property protected. The shorter that the lead time is between a forecast and a weather anomaly, the greater will be the constraint on society's ability to be proactive.

There are numerous examples in just about every country of *policies and laws* related to weather. These include but are not limited to the following:

- land-use regulations in flood plains;
- building codes in hurricane-prone areas;
- control of weather modification activities;
- hail, flood, lightning, and other weather-related insurance coverage;
- laws to encourage tall smokestack construction and, later, laws to prohibit their use;
- regulations of the location and emissions of air polluting industries;
- wind shear and policies affecting aircraft safety;
- severe weather warnings; and
- air pollution alerts.

A major problem, however, relates to the enforcement of these policies and laws. It was learned, for example, after Hurricane Andrew destroyed a large number of homes in southern Florida in August 1992 that the building standards had neither been met by the construction companies nor enforced by the housing authorities prior to the onset of Andrew.

Early warnings for specific weather and climate extremes are a major goal of those involved with the operational aspects of weather. The following list of anomalous weather events notes the lead times that forecasters can provide.

- Wind shear (hours, minutes)
- Hail (hours, minutes)
- Tornadoes (hours, minutes)

- Heat waves (days, hours)
- Frost (days, hours)
- Snowstorms (days, hours)
- Tsunamis (days, hours)

By contrast, early warnings about some weather and climate extremes provide societies with a longer lead time to act.

- Hurricanes (weeks, days, hours)
- El Niño events (years, seasons, months)

What may prove to be a useful early warning to spark preparations for some extremes may prove to be too early a warning for some weather anomalies and too late for others. Therefore, early warning systems are tailored to the specific type of weather extreme.

Weather policy and law, or the absence thereof, result from *weather politics*. Just about every weather impact on society at any spatial scale would provide an interesting story about the political aspects of weather. A few come to mind:

- Recent hurricanes Hugo, Andrew, Floyd, Georges, Mitch, and Michelle, among others, generated political issues that pitted politicians against each other. Interesting case studies are generated by hurricanes each year in the Atlantic and by typhoons or cyclones in the Pacific and Indian oceans.
- The Rapid City, South Dakota flood in the late 1970s was blamed on cloud-seeding activities that preceded the heavy rains in the region. Cloud seeders denied in court that their activities produced the floods.
- The Challenger space shuttle launch took place when temperatures were at critically cool levels, as noted by the engineers. Their feelings that the launch should wait for warmer weather were overruled by the launch managers, who put a higher priority on getting the launch off on schedule than on safety.
- Some states have accused other states of "cloud rustling." This is a term used by U.S. states that are downwind of cloud-seeding activities to produce snowfall during

wintertime droughts (e.g., Idaho against Washington; Colorado against Utah).

- Construction in low-lying coastal areas captures attention whenever a hurricane makes landfall and destroys beach-front homes, as happened during Hurricane Hugo in 1989. Nevertheless, such homes are continually being built in harm's way.
- The ongoing politics of providing airports and/or aircraft with the latest technologies to detect adverse weather conditions generated politicking by various members of the U.S. Congress, who sought to have the latest experimental technologies installed in their districts' airports.

Weather economics encompasses assessments of the value of weather information; the value of forecasts for various extremes; cost-benefit analyses of missed forecasts; the cost of the impacts of weather anomalies; and risk assessments. Figure 4.14 depicts the major weather disasters that occurred in the United States between 1980 and 2001. It supports the argument that the application of weather research findings to societal needs can be of great financial benefit to all levels of society.

Weather integrates ethical and equity issues into the atmospheric science research and research application agendas. Ethical and equity issues include questions such as:

- Are the poor (defined as those without political access) disproportionately affected by adverse extremes of weather, such as tornadoes and flash flooding?
- Do the poor have the same access to weather information as do the wealthier segments of society?
- Do the poor tend to be located in areas at higher risk to extreme meteorological events?

Weather technology is an important aspect of weather affairs. The development of engineering devices for weather detection purposes allows advances in the operational use of weather information in weather research. Doppler radar and downburst detection are just two examples of breakthroughs in technology and techniques that improve our understanding of weather phenomena.

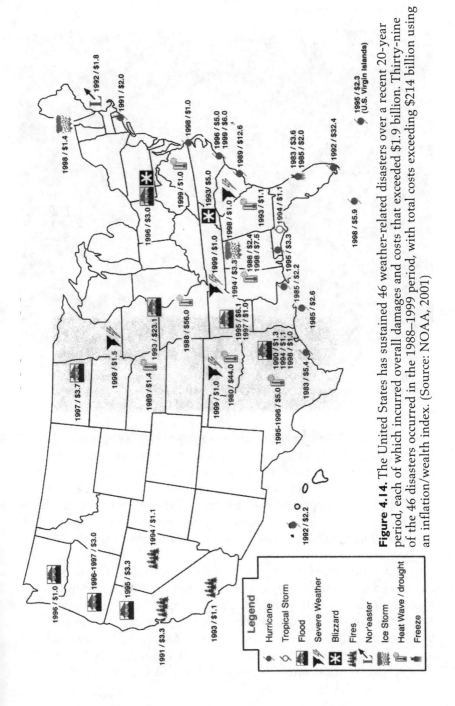

Figure 4.14. The United States has sustained 46 weather-related disasters over a recent 20-year period, each of which incurred overall damages and costs that exceeded $1.9 billion. Thirty-nine of the 46 disasters occurred in the 1988–1999 period, with total costs exceeding $214 billion using an inflation/wealth index. (Source: NOAA, 2001)

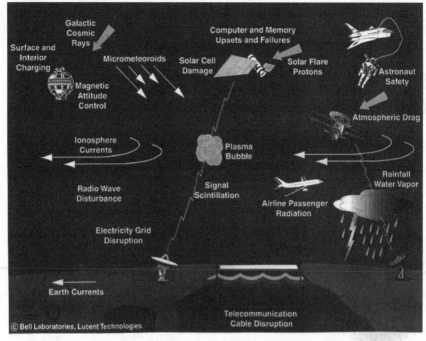

Figure 4.15. A variety of ways that storms in space can affect human activities. (Courtesy of Lou Lanzerotti, Bell Laboratories)

Space Weather (a.k.a. Storms in Space)

Space weather is worthy of the attention of present and future generations of students, decision makers, and the media, whether or not they have a strong interest in science. This becomes very apparent as one comes to realize just how much every person's life is influenced by the sun and solar activity. These forces influence our lives well beyond the sunburns we have come to fear.

Societies worldwide are rapidly becoming increasingly dependent on technological instruments for travel, communications, and monitoring of environmental changes on the planet. As a result of this dependence, they are increasingly becoming more vulnerable to variations in space weather. Figure 4.15 suggests some ways that human activities are affected by what one researcher has called "storms in space" (Freeman, 2001).

A U.S. National Academy of Sciences report defined space weather as "the conditions in space that affect Earth and its technological systems. Our space weather is a consequence of the behavior of the sun, the nature of Earth's magnet field and atmosphere, and our location in the solar system" (NAS, 1999). Freeman has noted:

> Just as violent storms that range in the Earth's lower atmosphere are manifestations of tropospheric weather, so also giant storms that take place in space surrounding the Earth are manifestations of what has become called "space weather." As reported by NOAA's Space Environment Center, space weather occurs in the area between the Earth and the Sun and refers to the disturbances and storms that swirl through space. . . . Much of what happens concerning space weather is dictated by the Sun. Solar flares, coronal mass ejections (CMEs) and other solar activities cause the stormy weather of space. (p. 13; see also NRL, 2003)

In a write-up about the mechanisms linked to disturbances in space, scientists at the National Center for Atmospheric Research (NCAR) succinctly described the process as follows:

> The sun is flinging 1 million tons of matter out into space every second. If you add all this up over the course of a day, it's comparable to the mass of Utah's Great Salt Lake. And this happens every day, day after day, year after year. This mass loss is called the solar wind. The solar wind is formed as the sun's topmost layer blows off into space carrying with it magnetic fields still attached to the sun. Gusts and disturbances form in the solar wind associated with violent events on the sun. The gusting and blowing of the solar wind against the Earth's protective magnetic shield in space is responsible for space weather storms. (Windows to the Universe, 2002)

Space weather forecasts can provide users of such information with lead times on the order of hours to a couple of days before the specific disruptions about which they are concerned are likely to occur. Disruptions may include excessive radiation exposure to airline passengers and astronauts, communication interruptions, or electric power problems.

As noted by the Space Environment Center, large geomagnetic storms are relatively infrequent, but when they occur, they can stress the susceptible systems for prolonged periods of time over large geographic areas. Secure operation of systems can still be maintained and hazards minimized if the occurrence, duration, and severity of impending storms can be accurately predicted in a timely manner. Thus, space weather forecasting is important for protecting national assets in both the commercial and military sectors.

At a recent meeting to develop NOAA's strategic plan for the first decade of the new century, space weather researchers and forecasters played a prominent role, so much so that by the end of the day, organizers had given space weather top billing along with weather and climate concerns. Perhaps, when looking toward the future, space weather will be a potentially important area of usable science. Although this is an especially valid view from the vantage point of a high-tech society, it is considerably less important to a large portion of the world's population.

Many of the world's billions are focused on trying to feed their families and to achieve a decent living. To get to an economic level of interest in or dependence on space weather, societies most likely will have to develop their economies to levels that provide them with the luxury of space weather concerns. To reach these levels of development, societies can start now to improve their understanding of climate affairs, an integral part of which is the more effective use of existing climate information.

Chapter 5 provides several examples of the ways that climate and weather information, including forecasts, have been used in a range of decisionmaking contexts. The political and socio-economic settings for the use of climate information are also discussed with a focus on climate-sensitive human activities. Examples are drawn from recent years, as well as from other decades. The examples encompass misuses of forecasts as well as success stories. Early warnings are central to the use of climate information because having this information in hand can empower the user. It also provides those who know how to use that information in their climate-sensitive activities the opportunity to avoid being surprised.

FIVE

USE OF CLIMATE INFORMATION IN DECISION MAKING

Climate affairs is a multidimensional field. Yet its many component parts share at least one common aim: the avoidance of unwelcome surprises, or rather surprises for which societies are unprepared. Avoiding surprise is no easy task, of course, because climate is characterized by chaos, instability, discontinuity, and anomaly. Thus, understanding climate is fraught with uncertainty, risk, and vulnerability, not to mention the bias on the part of researchers and policy makers.

The very notion of surprise is itself a subjective matter. There are events whose probabilities can be calculated and some events, although rare, that can in some circumstances be anticipated. In other words, there are knowable surprises, seemingly an oxymoron. There are also surprises in the course of scientific research, such as the discovery of the Antarctic ozone hole, and the realization that CFCs are ozone-eaters in the stratosphere, or the occurrence of the Tunguska meteor fall in the early 1900s (Kondratyev, 1988). Each of these was totally unexpected and, though now a feature of the historical record available to all living generations, it was previously unknown (see box 5.1). But even contemporary societies have not yet witnessed directly or indirectly the full range of climate variations, so they are not in a position to know about all the possible changes that the global climate system can or might go through, let alone what the impacts of those changes on regional climates might be.

Surprises can result because individuals and societies have short memories when it comes to previous climate-related anomalies and their impacts. Surprises can also result because of the relatively short time over which societies have been collecting reliable data and directly monitoring the climate system. Take, for example, the El Niño phenomenon. Following each event since the early 1970s, forecasters and researchers appeared to believe that they had a better understanding of the phenomenon. This led them to believe that the next event could be forecast with more accuracy. Unfortunately, those hopes would be dashed with each successive El Niño event because each El Niño behaved somewhat differently than its predecessors: it started earlier than anticipated, developed faster than thought possible, decayed very rapidly, was more intense than believed possible, and so on. Based on past experience, it is not unreasonable to expect that researchers will continue to be surprised, until they have witnessed enough events to identify the various combinations and permutations that the factors that go into making up an El Niño can take. Researchers were clearly taken aback when two very intense El Niño "events of the century" occurred only fifteen years apart (1982–1983 and 1997–1998).

Societies and researchers should expect climate-related surprises in the future at just about all locations on the earth. Although some surprises can be anticipated, others will truly be unknowable. It would be really surprising if there were no climate-related surprises in our future. The avoidance or mitigation of surprises is, nonetheless, one of the major underlying aims of the field of climate affairs.

Societies have prepared for and responded to climate variability, change, and extremes in many ways to mitigate, if not eradicate, surprise from their future. Measures have included attempts to shelter human activities from the vagaries of the climate system through "weather- and climate-proofing"; identifying ways to make and use climate and weather forecasts; setting up early warning systems; and most recently, treating climate as an

BOX 5.1
Stratospheric Ozone: A Lucky Escape

The development of the ozone hole was an unforeseen and unintended consequence of widespread use of chlorofluorohydrocarbons, as aerosols in spray cans, solvents, refrigerants, and as foaming agents. Had, inadvertently, bromofluorocarbons been used instead, the result could have been catastrophic. In terms of function as a refrigerant or insulator, bromofluorocarbons are as effective as chlorofluorocarbons. However, on an atom-for-atom basis, bromine is about 100 times more effective at destroying ozone than is chlorine. As Nobel Laureate Paul Crutzen has written, "This brings up the nightmarish thought that if the chemical industry had developed organobromine compounds instead of CFCs—or, alternatively, if chlorine chemistry would have run more like that of bromine—then without any preparedness, we would have been faced with a catastrophic ozone hole everywhere and at all seasons during the 1970s, probably before the atmospheric chemists had developed the necessary knowledge to identify the problem and the appropriate techniques for the necessary critical measurements. Noting that nobody had given any thought to the atmospheric consequences of the release of Cl or Br before 1974, I can only conclude that mankind has been extremely lucky" (Crutzen, 1995).

environmental and political security matter. Each of these measures is briefly discussed in the following sections.

Weather- and Climate-Proofing

Various governments over the years have proposed programs to weatherproof or climate-proof vulnerable regions within their countries. The objectives of such programs could be interpreted in either of two ways: one, to insulate human activities from the influence of potentially adverse weather and climate conditions, most likely extremes; and two, to reduce by any degree possible the exposure of weather- and climate-sensitive activities to climate-related hazards. The first objective is quite idealistic and can be misleading to the public because such a goal may be unattainable. To date, no society has been fully able to insulate its people and human activities from climate- and weather-related anomalies. Yet, a phrase such as "climate-proofing" suggests that there are programs in place that can now or, in the near future, achieve such an objective.

The second objective is more realistic, in that it suggests that climate-proofing is a process and that the steps toward achieving such a goal are doable and worthwhile. This objective is most likely the one that government bureaucrats have in mind when they propose "proofing" activities, either for society as a whole or for specific climate-sensitive social and economic sectors. Both objectives are geared toward minimizing surprises while at the same time mitigating the unwanted consequences of anomalous weather or climate.

For example, the history of agriculture in the Canadian prairies is filled with drought episodes. The prairie provinces suffered as much as the American Midwest during the Dust Bowl of the 1930s. In the 1970s, following severe drought in the Canadian prairies, the government launched a program to drought-proof the prairies (CDREE, 1978). Drought-proofing measures included changes in land-use practices, such as leaving stubble and crop residue in the ground after harvest. Expectations for successfully drought-proofing this region, however, were soon undermined by nature, as drought and crop losses continued unabated. Today, Canadian officials in the region are more specific in their attempts to protect farmers from drought impacts by, for example, calling for the drought-proofing of farm water supplies.

Despite the confusion in meaning that surrounds the concept of drought-proofing, it is still being proposed by UN agencies and governments. Two recent examples come to mind, Australia and India. During the 2002 drought in New South Wales, the government pursued a drought-proofing strategy, calling on farmers to review the way they manage their land and water resources for drought. Drought-proofing, in this situation, means mitigating the potentially adverse impacts of poor rainfall conditions by devising ways to keep moisture in the soil, by using no-till practices and by upgrading irrigation facilities (see, for example, Earthbeat, 2002).

As a second example, the UN Development Program has partnered with British and Australian development agencies in starting drought-proofing activities in India on an experimental basis (e.g., Orissa and Rajastan). The plan is to encourage the use of technologies that harvest rainwater and recharge groundwater to make water supplies in the rural areas more reliable and available than they are at present, especially in conditions of meteorological drought (UNDP, 2001).

Not everyone has bought into the notion of climate-proofing. For example, Indian policy analyst Devinder Sharma has argued that drought-proofing measures should not be imported from other countries but instead should be homegrown. In September 2002, he suggested the following:

> It comes as a rude shock. The American agriculture that we studied in the universities and appreciated has crumbled with one year of severe drought. It is well known that Indian agriculture falters because of its complete dependence on monsoons. But with the kind of industrialization that took place in the United States, and with the amount of investments made, we were told that US agriculture is not dependent on rains. Now, though, the drought-proofing that we heard so much about appears to be a big farce. (Sharma, 2002)

The U.S. government has also attempted weatherproofing. In late 1999, the U.S. National Weather Service launched a national computing system for forecasting. Within a matter of days, however, a forecast of light snow for the Washington, D.C. area proved wrong; a major winter storm developed, depositing twelve inches of snow in the metropolitan area (Pielke and Sarewitz, 2000).

Recently, in March 2001, a storm of major proportions—referred to by some forecasters as a potential storm of the century—was forecast for the lower half of the northeastern United States, including Washington, D.C. It was forecast to be a "nor'easter," the magnitude of which had not been seen since the 1950s. The forecast prompted people in the region to prepare for several days of coping with projected snow-related disruptions. Hardware stores were emptied of shovels, salt, mechanical snow-removing devices, and the like. Although a major storm did develop, its track unexpectedly shifted more than one hundred miles to the north. Most of the snow-related disruptions failed to occur in the Washington area. Once the storm had passed, the governor of New Jersey threatened to sue the National Weather Service for the costly impacts of what he viewed as a grossly erroneous forecast (McFadden, 2001).

Although labeling a program as climate-proofing or weather-proofing represents the hopes of the forecast communities, it is a poor way to capture the attention of the public. First of all, the notion can be interpreted to mean that such a goal is attainable with the availability of new forecasting tools and techniques and an improved understanding of the workings of the climate system. Second, it raises false hopes, which are only dashed to pieces by the next surprising weather or climate anomaly. A forecast is just a forecast. It does not come with a guarantee. Instead, it comes with an implied buyer beware label.

Climate Information

The development of methods to produce, on a routine basis, reliable and credible information about the future state of the atmosphere (and climate) would most likely merit a Nobel Prize. It is a constant desire of humankind to find ways to produce perfect information about the future state of the atmosphere and climate. Yet, many scientific uncertainties in our understanding about the climate system remain. Perfect forecasts are not to be. Improved forecasts? Yes. Perfect ones? No.

Today, researchers use sophisticated statistical measures as well as computer models, historical records, paleodata, proxy data, and even anecdotal accounts to reduce weather- and climate-related uncertainties with the hope of gaining a better glimpse of what future climates could be like.

Notwithstanding the practical limitations on forecasting the future state of the atmosphere and climate, considerable opportunity exists for the effective use of weather and climate information, which includes much more than just forecasts. Socioeconomic and political decision makers around the globe have increasingly become convinced that more information about climate, as uncertain as it might be, could play an important role in their decisionmaking processes. After all, they are forced to make decisions every day, usually without having the luxury of perfect information in hand. Only a few socioeconomic sectors have been repeatedly identified as especially weather- and climate-sensitive: agriculture, energy, water supply, health, and public safety. However, it is easy to show that other sectors of society are equally as sensitive to climate variability, change, and extremes: transportation, urban planning, tourism, economic development, clothes manufacturing, and the stock market.

Climate information, broadly defined, refers to any information that has a direct or indirect connection to atmospheric processes and to the structure, function, and impacts of the climate system. That includes

- information about the physical aspects of the climate system and its subcomponents and weather and climate variables such as temperature, precipitation, relative humidity, and albedo;
- historical information and proxy data (e.g., pollen samples in ancient soils, ice cores, tree rings) about climate that can yield insights into past climates;
- various aspects of ocean processes, levels, temperatures, thermal structure, and salinity, as well as information about ocean currents and coastal upwelling processes;
- climate's impacts on society and socioeconomic and political responses to climate variability, changes, extremes, and seasonal change; and
- ethical and equity perspectives of climate-environment-society interactions.

The general public is most familiar with weather and climate because of what's going on where they live and what they have personally witnessed, what they read about climate or hear from friends, and what forecasts they get from various media. Although much of the public is apparently skeptical about the reliability of such weather- and climate-related forecasts, they still seem to depend on them for information, entertainment, or just psychological comfort. Deep down, they know that someone is trying to give them a reliable, scientifically based glimpse of the future. Specialists in weather- or climate-sensitive activities, however, take those forecasts much more seriously.

The Greek philosopher Heraclitis once wrote "you cannot put your foot in the same river twice" because its flow is constantly changing. The same can be said of climate's impact on a society. Two droughts or hurricanes of the same intensity occurring in the same place but a couple of decades apart would most likely have different levels of impact. Increased impacts could be expected because of demographic changes, resulting in people living in different places, doing different things, and being dependent on each other in ways that differ from earlier times. For example, a hurricane that passes through southern Florida today of the same intensity as one that passed through the same area a few decades earlier is likely to have a greater economic impact because more structures have been built along Florida's hurricane-vulnerable coast (Pielke, 1995). However, the death toll would likely be reduced because society would have learned from the landfall of previous tropical storms about the need for and methods to protect people against hurricane winds, rains, and storm surges.

Today, reliable, satellite-dependent observation systems provide instant knowledge about the movement and intensification of tropical depressions, tropical storms, and threatening hurricanes. Forecasters provide approximate times and locations of hurricane landfall, constantly refining their projections as the hurricane moves closer. Local governments have established evacuation routes and hope to receive early warnings that will provide enough lead time to get large numbers of those threatened out of harm's way.

Historical accounts of weather- and climate-related impacts and societal responses to them can be very instructive to those who want to improve the way that societies cope with variability within the climate system. Once a significant number of case studies have been collected, reviewed, and evaluated, credible generalizations can be made. Although generalizations are not likely to match exactly any specific event, they can provide decision makers with usable information.

Cases can focus on the same location over time or can study

different locations at the same time. They can be compared for similarities in impacts of and responses to the same anomaly in different socioeconomic sectors. Identifying existing strengths and weaknesses in a coping mechanism can yield valuable information for decision makers who want to reduce vulnerability to climate-related hazards and to increase societal resilience to those hazards in the future. One can only hope that a society or a government that had once failed to protect its citizens from a climate-related disaster would want to avoid a similar failure in the future. Based on this premise, the adage "once burned, twice shy" was used as the basis for a UN-sponsored sixteen-country impacts and response strategies assessment centered on the 1997–1998 El Niño (Glantz, 2001b).

Not all successful decisions in climate- and weather-sensitive sectors depend heavily on climate information. Sometimes such information will prove to be useful to decision makers and at other times not. Nevertheless, such information should be taken into consideration. Even though, for example, forecasts of the development of El Niño events contain useful information for a wide range of decision makers, those forecasts are often not used. As a result, many policy makers around the world must react to El Niño's impacts rather than pro-act in response to the forecasts of such potentially devastating events.

An important aspect of the use of climate information relates to who has it and when. Forecasts can unwittingly perpetuate, if not increase, the gap between rich and poor in a given society (Pfaff et al., 1999). In most societies, the availability of timely information about climate anomalies can generate power and wealth, by allowing those with access to prepare better and earlier. Information is power. In a larger global context, countries that can afford to support a strong research establishment and an active science application program (e.g., research applied to national needs) have more usable climate information available than those that cannot. This translates into a matter of industrialized societies versus developing ones, or the North-South divide.

Wealthier societies are investing heavily in the forecasting of climate on several time scales. The forecasts with which people are most familiar are the most frequent ones, on daily or weekly time scales, but government agencies and various economic sectors are very interested in producing seasonal and interannual forecasts. They want to enhance the accuracy and value of forecasts by exploring new ways to expand their use.

For example, improvements in monitoring and forecasting El Niño and La Niña events are of major interest to the scientific community and to decision makers. Each of these extremes of the so-called ENSO cycle seems to spawn a corresponding set of extreme meteorological events. This chain of events in the physical aspects of the climate system provides decision makers with additional lead time for effective action. Researchers hope that one day these forecasts will provide many potential users with the earliest possible warning of potential climate-related problems.

Forecasting any departure from the normal range of climate conditions is a challenging task. The recent experience of more than a dozen forecast groups that tried but failed to forecast the onset of the 1997–1998 El Niño suggests that forecasters have a ways to go before they can reliably predict El Niño's onset (Barnston et al., 1999). However, once an El Niño process is known to have begun, forecasters have a better chance of predicting its potential impacts, especially in locations such as Peru, Ecuador, Australia, North America, southern Africa, the Pacific islands, and Southeast Asia.

Knowing that there is a reliable correlation between El Niño and, say, flooding in northern Peru enables a government with fair warning of El Niño's onset to clear dry river channels, reinforce bridges, protect against possible mudslides, alert the population about flooding and about water-borne disease outbreak possibilities, and so on. Pacific island governments can prepare for El Niño-related droughts by enacting various water harvesting and water-saving techniques to fill up local reservoirs. Awareness of a potential climate-related problem is a first step toward preventive decision making.

Retailing and Wholesaling Climate Information

There is little doubt that throughout the 1990s many decision makers and the public became increasingly aware of the potential value in the use of El Niño information. Even without a full understanding of the science of El Niño, they were made aware of its existence, and many people developed at least a vague idea about its possible impacts. The 1997–1998 event was in fact the second wake-up call about El Niño-related climate problems; the first call came with the 1982–1983 event. However, awareness of El Niño and forecasts about it do not by themselves translate into usable scientific information. What is needed are science information brokers positioned between the information producers and potential users. Brokers would serve to translate the scientific information into a user-friendly language and explain to potential users how best it might be used.

An important distinction needs to be made between "wholesaling" climate information and "retailing" it. Wholesaling can make people aware of a phenomenon and its impacts in a general way. Although it may be interesting information to the general public, in most cases it is not usable by those who do not really understand it or its true implications for their daily activities or for their societies. General statements about an impending El Niño do not contain enough of the timely, reliable detail needed for tactical decision making. For most people, corporations, and government agencies, it is necessary to "retail" climate information, tailoring it to the specific needs of specific users.

More generally, climate information placed in the right hands at the right time in an easily understandable form and with appropriate lead times to address the task at hand can prove to be a valuable decisionmaking resource.

Successful Uses of Climate Information: Some Examples

California's Use of the 1997–1998 El Niño Forecast

The mature phase of an intense El Niño was forecast in mid-June 1997 to occur in late fall and early winter. For some areas, such as the West coast of the United States, a June forecast of El Niño provides about six months of lead time to prepare for its earliest impacts. Many Californians did just that: they took the forecast seriously, having been reminded of the devastating floods related to the intense 1982–1983 El Niño. At that time, there was no forecast; forecasters were caught by surprise with the appearance of the first "El Niño of the Century."

In 1997, riverbeds were cleared of debris, as were the open flood drains and sewers, to let the expected voluminous runoff from the heavy rains flow unobstructed to the sea. This helped to minimize the risk of flooding in many areas. Advertisements appeared in the newspapers and on TV calling on people to fix their roofs before the rains came.

California's Senator Barbara Boxer called for a public meeting on how to cope with the impacts of the upcoming El Niño event. The summit was held in mid-October 1997 and was sponsored by the U.S. Federal Emergency Management Agency and the State of California. Vice President Al Gore attended. He used the occasion, which was two months before COP3 in Kyoto, to assert that El Niño's intensity and frequency would be enhanced by global warming.

Some climate modelers and forecasters projected heavy rainfall (on the order of three to four times the long-term average) for southern California, while many others downplayed such an apocalyptic warning from the summit, various forecasters, and the media. When the heavy rains failed to materialize in November and December and well into January 1998, various groups and individuals began to joke about the forecast and the forecasters. Headlines such as "El No-Show" appeared. An El Niño website called "El Hypo" was developed.

All that ridicule ended, however, when heavy rains and severe coastal storms arrived in southern California in late January. TV commentators dedicated major segments of their news shows to coverage of El Niño and its global impacts. The forecasters were finally vindicated (Nick Graham of Scripps Oceanographic Institute was right!) as El Niño-related coastal storms eroded beaches and the base of cliffs, causing homes to topple into the Pacific Ocean.

The State of California engaged several of its bureaucratic agencies in El Niño-related activities in late 1997 in its attempt to prepare for El Niño's impacts. The list of agencies is interesting because it includes some that even climate impacts experts might not have taken into consideration (e.g., the last one on the list).

- Governor's Office of Emergency Services
- Department of Water Resources

- California Resources Agency
- Business, Transportation & Housing Agency
- Employment Development Department
- California Environmental Protection Agency
- Department of Finance
- California Department of Food & Agriculture
- California National Guard
- California Public Utilities Commission
- Youth & Adult Correctional Agency

California's response to El Niño's potential impacts is very instructive. Most important was the coming together of several government agencies, many of which on the surface would seem to have little to do directly with El Niño or even with climate anomalies. A key lesson from the California experience is that all government agencies must improve their level of awareness of how climate anomalies can indirectly affect their operations. Forewarned of potential climate-related ripples throughout society, economic planners can prepare agencies to take actions against the potential impacts.

Palm Oil Futures: The Philippines and the Ivory Coast

In the mid-1980s, two researchers from a West Germany corporation paid a visit to my office. They were seeking information about El Niño and its possible influences on agricultural production in general, or so they said. I noticed a subtle bias toward Southeast Asia in their questions. I repeatedly asked what they were specifically after. Keep in mind that up to that time there had been relatively little interest in the science of El Niño. After evading my questions for a while, they admitted that as representatives of a major soap-manufacturing corporation, they wanted specific information on the impacts of El Niño on palm oil production in the southern Philippines. Palm oil was and still is an important oil seed traded on the futures market. Buying futures contracts at too high or too low a price would have major implications for traders and corporations that rely on this commodity for their exports. If palm oil production declined because of drought in the southern Philippines, then the price would rise. Those who sold palm oil futures would then be legally bound to fulfill the orders at the predetermined price, which was based on the assumption that normal weather conditions would prevail.

If the connection between El Niño and a reduction in palm oil production could be reliably established, those who use such information wisely could fare well financially. Also, if palm oil production is down in Southeast Asia, for example, it could be up in other parts, such as the Ivory Coast. Hence, a corporation with palm oil plantations in different parts of the globe could reduce the impacts of climate variability in palm oil supply and price in the international marketplace. The two researchers were attempting to use specific climate information as a resource to allow this corporation to enhance its profits. In addition, they were interested not only in climate conditions where their corporation had facilities but also in regions where their competitors were growing the same commodity for export.

Misuses of Climate Information (Including Forecasts)

Indonesian Forest Fires and El Niño

In 1997 and 1998, considerable attention was drawn to the forest fires in Indonesia. Fires burned across more than nine million hectares. The fires were initially blamed on El Niño-related drought. The haze that emanated from the fires and dispersed throughout the region captured global attention. The geographic extent of the problem was captured by satellite images. Similar but less extensive fires occurred during the 1982–1983 El Niño, destroying an estimated three million hectares. Those fires, too, had been blamed on the El Niño-related drought in Indonesia and Malaysia. To be sure, when an El Niño occurs, there is a high probability of drought throughout Southeast Asia in general and Indonesia specifically.

The Indonesian government in power in each of these two periods blamed El Niño for the severity of the fires and the damage to the country's tropical rainforests. However, journalists and researchers alike discovered that many company directors and entrepreneurs had hired peasants to set the fires because the existing rainforest could not be destroyed legally to convert the land to any other use. However, if the forest burned due to natural causes, such as drought or El Niño, then the land could legally be converted into large-scale plantations and agricultural plots. The awareness that the rainforests and soils are usually dry during an El Niño event encouraged the entrepreneurs (many of who were financial partners with high-ranking government officials) to set large tracts of the rainforests on fire and then to sit back and blame El Niño. More than twenty-nine companies were cited for violating the country's laws to protect the rainforest.

Forecasts of the Caspian Sea Level

The Caspian Sea, the world's largest inland sea, has no natural outlet to the ocean. It is fed mainly (80 percent) by flow from the Volga River system. Over geological time, the sea's levels have varied greatly as the result of natural variations in the global climate system and precipitation in the Volga basin. In the last two hundred years, the level of the sea has risen and fallen by a couple of meters in either direction (Voropayev, 1997).

Between 1840 and 1930, the sea level was relatively stable. However, in the early 1930s, the level began to drop and continued to do so until 1977. In 1929, the level was measured at –25.9 meters (the Caspian is below sea level), in 1956 at –28.4 m, and in 1977 at –29 m. According to a Russian geographical encyclopedia (Zonn, 2003), that was the lowest level of the sea in 400 years.

During this period of sea level drop, the Soviet government became concerned. Discussion about the possibility of the Caspian permanently drying up sparked the idea of transferring water from the Soviet Union's northward-flowing rivers toward the Caspian to stabilize the sea level, to which settlements around the shoreline had become accustomed. In the end, this was not done, although the idea remains alive. Other actions were taken, however, such as the damming of the Karabogazgol, a large bay in Turkmenistan along the eastern coast of the Caspian. This would stop water from the sea from flowing into the bay. Settlements and various human activities followed the receding shoreline because the Soviet government, if not the country's researchers, felt that the desiccation of the sea would be a long-term process, if not a permanent one. Their projections proved to be wrong.

In 1978, to the surprise of most people, the Caspian sea level began to rise and continued to do so for the next couple of decades. The constantly rising sea destroyed settlements, farmland, and infrastructure in the Soviet Union and Iran that had recently been built on the receding shoreline. Between 1978 and 1995, the sea level rose by about 2.5 m. New sea level projections emerged, based on computer modeling efforts of Russian climatologists and hydrologists, suggesting that the level would likely continue to rise until about the year 2015. However, following 1995, that projection was proven wrong. The level dropped some tens of centimeters and has stabilized over the past few years.

Reasons for the fluctuations in sea level have been suggested: a decadally fluctuating climate, economic activity involving reservoir and dam construction in the upstream and midstream regions of the Volga basin, or both. Aside from the influences of regional climate, a major impact on the sea's level was the numerous hydrological facilities constructed along the Volga River beginning in the 1930s. Water was taken from the river to fill reservoirs to capacity in the next few decades, thereby depriving the sea of this volume of water.

The Caspian situation shows how important it is to improve the analytical assessment tools for separating climate-related effects from other factors when trying to ascertain the impacts of climate on society. This can help government officials to apply scarce resources and attention to the true causes of their climate-related problems, as opposed to apparent and misleading ones.

The 1977 Yakima, Washington Drought

Several droughts occurred in 1976 in various parts of the western United States. However, drought had not yet affected the state of Washington. Nevertheless, water resource managers in the Yakima River basin, reading about droughts in areas around the Yakima watershed, had convinced themselves that a drought was sure to spread to their region. They ran their streamflow models, as they had done for many years, to identify the expected level of streamflow during the 1977 summer irrigation season. The output of their model runs suggested that streamflow would be adequate to meet the needs of all users, those with junior as well as with senior water rights. In times of low streamflow, those with senior water rights, that is, those who had first used river water for irrigation, were to get their share of water first. Those with junior rights arrived in the western United States later in time. The rule of thumb that captures water law in the western United States is "first in time, first in right."

However, the U.S. Bureau of Reclamation's water managers in the Yakima River basin did not believe the results of their model runs and opted instead to reduce the amount of water available to be shared among irrigation farmers to one-third of the normal amount of streamflow. In January 1977, they decided to divide that amount among those users who had senior water rights, leaving but six percent of normal to be shared by those with junior water rights.

Within a couple months, however, forecasters and water managers as well as irrigation farmers could see that there was near-normal water in the river system and that the water was flowing unused past their fields on its way to the sea. Instead of revising their forecasts or revising their mandated water allocations, the Bureau's water managers waited for the drought conditions that they had forecast, but which had not yet appeared, to catch up with their forecasts. This water allocation decision proved to be very harmful to farmers with junior water rights. Some of these farmers chose to dig expensive wells to water their fields and livestock. Others decided not to grow annual crops that season at all. Still others sold off all their livestock for fear of being unable to provide them with adequate fodder in winter. One farmer even chose to move his entire mint crop from the Yakima River basin to the Columbia River basin. These and other costly actions were taken to minimize the adverse impacts of drought.

Making a bad situation worse was the fact that no drought materialized in the Yakima basin that year. In fact, there was enough water in the river to take care of the needs of all the farmers and livestock owners, regardless of water rights. Many farmers joined together to bring a $20 million lawsuit against the Bureau of Reclamation and the U.S. government. Eventually, the lawsuit was dropped because the U.S. government did not allow itself to be sued, which was its prerogative.

This case study exposes some problems associated with streamflow forecasts as well as with the perceptions of water resource managers about the output of the models they use to project seasonal stream flow patterns. In the Yakima case, the streamflow model in use had not been tested by nature—until 1977. The study also suggests that one incorrect forecast may cancel out the positive value of several good ones (Glantz, 1982).

Zimbabwe Drought, 1991–1992

In the late 1980s and early 1990s, the government of Zimbabwe was engaged in negotiations with the International Monetary Fund (IMF) about structural adjustment of the country's economy. In 1991, the government began to enact some of the measures proposed by the IMF, such as selling off its maize reserves. Zimbabwe, along with South Africa, had been a regional exporter of food. The IMF representatives argued that the country's reserves were well beyond that which was needed in a national food emergency. Besides, there had not been a major drought in the country for some years. By reducing the amount of stored maize, funds could be saved.

Unfortunately, the IMF apparently did not pay much attention to the poor harvest in the previous year (1990) and the already lowered national maize reserves. The IMF representatives had apparently also failed to take heed of the El Niño forecasts that had been issued by some research groups in 1990 (no event emerged) and again in 1991 (an event did emerge).

There is a strong statistical relationship between drought in southern Africa and sea surface temperature changes in the central and eastern Pacific Ocean. The relationship was identified as early as 1982; when an El Niño occurred, there was about an 80 percent chance that drought would occur in southern Africa (Rasmusson and Carpenter, 1982).

Drought in Zimbabwe causes a sharp reduction in food production at the national level, which in turn leads to higher food prices in the marketplace. Higher prices lead to a reduction in access to food for the poor. Another impact of drought is a reduction in grain exports to neighboring countries in sub-Saharan Africa. Thus, putting the known pieces of information together—El Niño's influence of southern Africa, drought, food security—one could have foreseen a possibility of a need to maintain food reserves. Unfortunately, those pieces were not put together, and drought and food shortages did occur. In response to poor food production, the government of Zimbabwe was forced to buy back its maize at prices three to four times those it had sold it for several months earlier (Betsill et al., 1997).

This case underscores the need for national governments and organizations, especially those involved in international humanitarian aid and financial assistance, to understand the extremes of the ENSO cycle and the distant impacts on socio-economic conditions worldwide.

The Colorado River Compact, 1922

In 1922, a legal arrangement for the sharing of water in the Colorado River system in the American West was developed between the states in the upper part of the Colorado River basin and those in the lower part. The compact's apportionment was based on streamflow data averaged from 1900 to 1920. It was decided that 7.5 million acre-feet per year on average over a ten-year period would be provided by the upper basin states to those in the lower basin. However, some years after the compact was signed, Colorado River streamflow began to drop and this continued for the next several decades. It was later realized that the compact's apportionment had been based on the wettest two decades in a few centuries (1900–1920). Instead of using percentages of available streamflow over years or decades to apportion water in the system, allocations were guaranteed in absolute amounts. Although Colorado streamflow data for the latter part of the 1800s had been available in the early part of the 1920s, they were not used when negotiators were calculating the absolute amount of water to be divided. As a result of this legal agreement, drought-related water shortages in the river system have to be absorbed by the upper basin states, because they had agreed to provide, under all circumstances, a fixed amount of water to lower basin states, primarily California.

Putting Forecasts in Political Context

Many scientists believe that the mere existence of a seasonal forecast based on computer model outputs, satellite images, and statistical measures, when placed on the Internet, will be of great benefit to societies around the globe. Yet, producing a reliable forecast or having one in hand may not be enough to ensure that societies will benefit from it. How a particular forecast is perceived (as reliable and credible or not) and the ability of a society to act on it determines its usability and value to decision makers.

A brief comparison of how three governments, Peru, Kenya, and Costa Rica, responded to the forecast of a strong El Niño by mid-June 1997 highlights the effect of different political contexts for forecasts of the same El Niño.

When Peruvians received early forecasts in April and May and a reliable one in June 1997 of a possible El Niño, Peruvian President Fujimori reacted promptly, establishing a high-level task force to prepare Peru for the likely impacts of the event. Fujimori then sought and received financial support from the World Bank to cover the cost of preventive action. This was the first time that this president took an El Niño forecast seriously, even though a few events had occurred earlier during his presidency. Some observers argue that he did so because he had finally learned of El Niño's damaging impacts on his country. Others argue more convincingly that he reacted quickly to the U.S. National Oceanic and Atmospheric Administration's forecast of a strong event (Peruvian scientists had forecast a weak event) to use his proactive response to El Niño as a way to increase his popularity. Fujimori had already decided to run for president for a third term, bypassing constitutional limitations on presidential tenure. TV coverage of the president personally combating El Niño's impacts throughout the country would help to boost his image among voters.

The Kenyan government received the forecast of a strong El Niño about the same time as did Peru from its national meteorological service, which had also been receiving forecasts from foreign sources. The government, with a reliable El Niño forecast in hand, took no action. Factors affecting the government's lack of response could be any one or a combination of the following:

- a perceived unreliable track record of seasonal forecasts by Kenyan forecasters;
- a weak or unconvincing statistical connection between anomalously warm sea surface temperatures thousands of miles away in the tropical Pacific and heavy rainfall in parts of Kenya; and
- a lack of financial resources within the country to take preventive actions.

Heavy rains and flooding did occur in November 1997, as had been forecast. Then the Kenyan government asked for emergency assistance from the World Bank, which did provide financial assistance, but with strings attached because of known widespread corruption in the government.

Costa Rica has a highly skilled national meteorological service. Its scientists are well informed and well linked to various climate research and operational forecast units in the United States. As a result, the service received El Niño forecasts through personal contacts with major meteorological centers around the globe, as well as by way of the Internet. Although reference to an impending El Niño first appeared in March 1997, the meteorological service director chose to wait a few months until confidence was stronger that an El Niño was in fact developing. The government requested that the service prepare a proposal for financial assistance from the World Bank, requesting funds to enable the financially strapped democratic government to prepare for the potential impacts of El Niño. The bank rejected its application, suggesting that Costa Rica could request assistance once the event had adversely impacted the country. The government then instructed its meteorological service not to play up the possibility of an El Niño event because it was not in a financial position to fund proactive measures. Perhaps government officials also hoped that the event's impacts might somehow bypass Costa Rica. As it turned out, drought, floods, and fires affected Costa Rica, at an estimated cost of $100 million.

This three-country comparison is as enlightening as it is interesting. It suggests that the existence of a reliable forecast for El Niño or, for that matter, any climate-related event does not guarantee that the most appropriate responses to it will be taken or will be taken for the right reasons. The comparison also suggests that responses to forecasts are not a function of the type of political system that receives them. Authoritarian as well as democratic regimes responded in unexpected ways. Thus, the type of political system in a country does not guarantee the way the government might respond to an El Niño forecast. This comparison also showed that governments were aware of the problems that might be generated by an El Niño thousands of kilometers away in the central Pacific but still failed to take actions in preparation for those impacts. Having information available in the right decisionmaking hands at the right moment does not guarantee that the information will be used to mitigate foreseeable impacts. Political and economic factors must also be taken into account.

These three country studies were part of a sixteen-country assessment—"Reducing the Impacts of Environmental Emergencies Through Early Warning and Preparedness: The Case of the 1997–1998 El Niño" (Glantz, 2001b). The assessment sought to identify how governments coped with both the forecast of an impending El Niño and its impacts. Among its key findings were the following:

- Many governments are aware of the climate-related hazards that their citizens face, but for a variety of different reasons, they fail to take the steps necessary to prepare for them.
- Each country affected by El Niño must create national expertise to understand the phenomenon and its potential impacts on its territory and on the territory of those countries with which it competes.
- El Niño forecasts provide policy makers with the earliest warning of possible climate-related problems.
- A need exists for societies worldwide to focus more directly on developing a national *culture of prevention* for climate and climate-related hazards, as opposed to defaulting to a *culture of adaptation*.

Using Climate Information for Early Warning: Climate-Related Hotspots and Flashpoints

The notions of hotspots and flashpoints have become popular in recent years. They are used to identify activities or changes in ecosystems that may lead to long-term environmental degradation. Climate is frequently a contributing factor in the occurrence of hotspots and flashpoints.

A search on the Internet uncovers a wide range of uses of the notion of hotspots: political and military conflict hotspots, sea surface temperature hotspots, biodiversity hotspots, coral bleaching hotspots, chaotic weather hotspots, and geological hotspots, among others. To the general public, a reference to hotspots will more than likely conjure up an image of a favorable experience. An Internet search also uncovers social hotspots: cultural hotspots, skiing hotspots, tourist hotspots, scuba hotspots, and so forth.

"Hotspots" is a strange notion, in that it has opposing meanings that depend on the message to be communicated. Its specific meaning depends on the adjectives and the contexts that are used to describe it. From the examples above, one can see that there are at least two general meanings of hotspots: a negative sense and a positive sense. The NGO Conservation International (CI), for example, has produced a map that identifies biological and soil hotspots. These are locations at risk of losing biodiversity (Conservation International, 2002). Some of these hotspots are linked. For example, in a UN field survey to identify coral hotspots, a press release noted that "eight of the ten coral reef hotspots are adjacent to a terrestrial biodiversity hotspot" (UNEP, 2002). CI also referred to "bright spots," or areas where soil degradation had been arrested or reversed.

Every ecosystem is at risk to changes resulting from external influences, natural or anthropogenic, or a mix of the two. Each environmental change, however slight, could eventually prove to have started the unraveling of an ecosystem's sustainability over the long term. Whether an uncontrolled unraveling eventually occurs often depends on decisions made by individuals and governments. In many cases, influences on the environment, especially those related to climate, cannot be avoided.

The growing demand for food, fiber, and cash crops, along with pressure on the land because of increasing populations, has led to an increasing number of spots around the globe where agricultural activities, climate variability, and environmental conditions have intersected in various ways. Farmers may encroach on forests to cultivate new areas, either by design or out of necessity. Local entrepreneurs may carve shrimp ponds out of the mangroves to engage in lucrative shrimp farming. Cultivators may convert grazing areas into farms for food or cash crop production. In many instances, human activities have migrated into areas that are marginal for these new agricultural or rangeland uses. Environmental degradation, soil impoverishment, and poverty often follow.

Most environmental changes are of the creeping kind, for example, slow onset, low-grade, incremental, but cumulative over time (Glantz, 1994a, 1999). Unfortunately, in such situations, thresholds of adverse changes usually become visible only after they have been crossed. Thus, thresholds of environmental changes (e.g., for land transformation and degradation) need to be determined subjectively in advance to allow time for preventive action.

Government agencies, each with their own jurisdiction and area of concern, are interested in the earliest credible warning of potential problems they may have to address. They are interested in identifying as far in advance as possible the factors that might add to cultural, environmental, or governmental instability. Identifying potential and existing hotspots can help to satisfy these interests. It is useful to put the notion of hotspots within a broader context, which can be represented schematically as a pyramid (figure 5.1).

The base of the pyramid represents the pristine environment, unmodified by human activities. The next level represents land transformation or a limited interference with the environment. There is not enough human activity to interrupt environmental processes; for example, land that is transformed to a limited extent from rangeland to farmland or from forest to rangeland or farmland. These changes are not necessarily forms of degradation of the land's surface or of the vegetative cover. These are changes in land use that can lead to an increase in the land's value

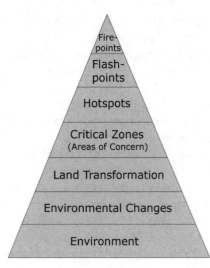

Figure 5.1. Flashpoints pyramid.

to society, depending on their geographic scale and the kind of land conversion. In a way, this level of change can be viewed as a transformation from one kind of activity to another within the same climate regime. The point is that not every interaction between agriculture and the environment is a negative or a zero-sum interaction, in which either the environment or agriculture wins and the other loses.

The next level of the pyramid represents critical zones or areas of concern. These are fragile zones, such as the tropical rainforests and arid lands, which are well known to be highly susceptible to degradation in the face of inappropriate land-use activities and/or a variable or changing climate. Areas that had been transformed for agricultural purposes often become the sources of critical zones, as agricultural activities intensify. Critical zones can encompass entire countries or just specific ecosystems within them.

Arid lands, for example, can be exploited sustainably using certain known land-use practices. However, the vegetative cover and soils of arid lands can be easily degraded if inappropriate practices are pursued. The irrigation of dry lands has been done sustainably in many places. However, in the absence of proper drainage facilities, salinization and waterlogging of the soils begins to occur but at this point, the adverse processes are still reversible. As a result, the salinized soils must be flushed repeatedly of toxic salts before new planting can take place. Eventually, the land has to be abandoned, as has been the case in the Aral Sea basin.

The next level of the pyramid represents hotspots, or locations where changes in land use or in the type of land used for agricultural activities are believed to impinge on ecological health. Hotspots are areas where reductions in soil fertility, declining crop yields, and so forth, have become more serious and border on becoming irreversible. Following on the above example, in the absence of proper drainage of irrigation water, soils become increasingly salinized. At first, this manifests itself with reduced crop yields, and eventually a salty crust appears at the surface, making the land unusable for agriculture. Hotspots can be further divided into hot, hotter, and hottest, depending on the degree or rate of degradation.

The next level represents "flashpoints." In chemistry, the flashpoint is the temperature at which a flame will ignite the vapors of a flammable liquid. When the flame is removed, the fire will go out. The social equivalent to a flashpoint would be a *jacquerie* (uprising), such as a bread riot because of food shortages. Once the government supplies bread to the people, the riot ends. The notion of *jacquerie* applies to climate and climate-related flashpoints. These are the last of the early warnings that a government will get before a crisis leads to a situation that is out of its control.

Flashpoints can appear as catalysts within a hotspots region. They can contribute to the instability of governments, economies, cultures, or ecosystems. Examples of climate-related destabilization of a government or an economy have involved droughts, fires, frosts, or cyclones. Second-order aspects of climate impacts include food or water shortages, forced migration, famine, and haze. Severe drought became a climate-related flashpoint that resulted in the fall of Emperor Haile Selassie's regime in Ethiopia in 1974 and caused problems for President Robert Mugabe's regime in Zimbabwe in late 2002. Most often, meteorological drought by itself is not the only problem facing a government at a given point in time. Thus, the source of instability is often drought plus another adverse environmental, socioeconomic or political change. For example, drought + high food prices can lead to severe food shortages, if not famine; drought + political instability can lead to a coup d'état; drought + poverty can lead to migration by the poor into marginal areas in search of farmland. This can be called the "drought +" factor. Several recent political instabilities have been linked to El Niño (Glantz, 2001b).

One specific example of a climate-related flashpoint occurred in North Korea in the late 1990s. A multiyear drought and an isolationist ideology combined to catalyze famine in North Korea and later played a key role in that government's first-ever efforts at rapprochement with South Korea, the United States, and other Western governments. The prolonged drought, combined with mismanagement of the country's agricultural sector and the delayed appeal for food aid, resulted in famine deaths among the people and in international embarrassment for North Korea (Crossette, 1999).

Flashpoints can be identified in at least two ways. Because climatologists have some skill in forecasting El Niño's teleconnected hazards in some areas, one way would be to identify regions of the globe that are likely to be affected by El Niño or La Niña. Then, within those regions one could identify the political or economic situations that are already under pressure for other reasons. Forecasts of these extremes can provide decision makers in several countries with the earliest warning of potential climate-related flashpoints. These forecasts, despite their level of scientific uncertainty, can provide useful lead time for tactical, if not strategic, decision making at various levels of society.

A second approach would be to identify countries that are already experiencing some degree of political, economic, or cultural instability and then identify within those countries chronic climate and climate-related problems, for example, droughts and floods, that could intensify the existing instability.

The very top of the pyramid represents "firepoints," the point at which unwanted permanent changes occur. It is the point at which the climate-related factor combines with other factors to topple a government or to generate conflict among protagonists. One could say it is the point of no return to earlier conditions, environmental or political. In chemistry, a firepoint is the temperature at which a fire will continue to burn even if the source of heating is removed. The social equivalent would be as follows: If the cause of a *jacquerie* is left unaddressed (e.g., no bread is given out), the uprising can turn into a full-blown revolution. For example, the eventual response by the North Korean government to request and accept international assistance averted a flashpoint situation from becoming a possible firepoint.

In sum, because no one has yet figured out how to forecast the future with any high degree of reliability, societies must rely on the educated guesses of people they believe to be knowledgeable. One way to get a glimpse of the near future is to focus on the recent past to identify how various aspects of climate might have destabilized a government, an economy, or an ecosystem.

The political, social, economic, and even cultural benefits of a hotspots approach to early warning might be:

1. an awareness of potential sources of instability;
2. an awareness of the locations in which they are likely to crop up;
3. a useful lead time in which to warn those at risk or to prepare responses;
4. a heads-up for other governments about the possibility of the spillover of instability across either a domestic political jurisdiction or across an international border; and
5. the provision of useful lead time for decision makers to address underlying socioeconomic conditions.

The list of hotspots susceptible to becoming either flashpoints or firepoints will most likely change from year to year, as a given country's underlying causes of instability may vary at a given point in time. Development economist Jeffrey Sachs recently commented on the importance of transitions with respect to economic development prospects. He suggested that "the main lesson about transitions is that small amounts of help at crucial moments can tip the balance toward successful outcomes" (Sachs et al., 2001).

Then there are *blind spots*. Policy makers are operating in a multistressed decisionmaking environment. They have lots of incoming information to deal with and are under pressure to make decisions quickly in favor of the various interest groups trying to influence them. They also have their own personal political, social, religious, or ideological biases to contend with. They are forced to pay selective attention to certain issues while being inattentive to others. As a result, there are blind spots in the political arena when it comes to climate-related and other kinds of environmental degradation. Even when certain issues are brought to their attention, the policy makers may show little interest, for any one of a variety of reasons:

- those affected have little political power;
- by the time the degradation reaches crisis proportions, the policy maker will be out of office;
- the degradation will become a crisis only in the distant future;
- the profits derived from the processes causing the degradation outweigh the perceived costs of the degradation;
- a belief that those who caused the degradation should take steps to stop it; and
- the ecosystem and its inhabitants are marginalized from a national standpoint.

Environmental Security

Environmental security is a growth research industry sparked by the end of the Cold War. Its original focus was on how environmental changes might lead to acute political conflict which involves a high probability of violence (Homer-Dixon, 1991). Since its inception, the concept has been broadened considerably to encompass a wide range of conflicts. Much like the notion of sustainable development, this idea too has taken on a variety of meanings. It is a controversial notion, in that it has supporters (e.g., Homer-Dixon, 1999) and detractors (e.g., Deudney, 1990). The phrase easily captures one's attention, but you must know who is using it to get at the meaning intended by its user.

Environmental security, in my view, relates to changes in the environment that can lead to instability in a region, a political system, or an economy. Environmental security is a reason for a hotspots or flashpoints approach to early warnings about possible future climate-related changes. Climate anomalies such as drought can be a destabilizing force, depending on how a government responds to them. As an example, there were four drought-related coups in Africa's Sahelian region during the first half of the 1970s. The perpetrators of the coups stated that their efforts were on behalf of the neglected drought and famine victims.

If a country's water supply were cut off by actions taken in another country, this could impinge on the security of the country deprived of water. The pollution of transboundary rivers could also serve as a cause célèbre for initiating conflict and perhaps generating insecurity. Converting fertile land into cash-crop production from subsistence farms could also lead to environmental insecurity. Any major shift in the availability of a formerly shared fish stock could lead to a conflict situation.

Soil erosion, per se, does not have to lead to destabilization of a government. However, widespread unfettered soil erosion that adversely impacts food production, and therefore food availability, could put considerable indirect pressure on the stability of the government in power.

The problem of environmental security arises over the kinds of environmental changes that one places under its umbrella. Deforestation, soil erosion, air pollution, declining water quality, and desertification processes do not necessarily impact the stability of a government unless other pressing issues are at play as well.

Homer-Dixon (1994), one of the people originally researching environmental security, has since stopped talking about security and instead talks of environmental scarcity (Homer-Dixon, 1999). Climate-related environmental security can be addressed through the hotspots/flashpoints assessment process.

Forecasts Are the Answer, But What Was the Question?

Societies everywhere and throughout time have sought to gain a glimpse of their futures. Today, we do it through computer modeling activities for a variety of processes—economic, political, social, cultural, and the workings of the global climate system. Perhaps it is part of human nature to want a glimpse of what lies ahead or perhaps it is a manufactured demand by some elements of society to get people looking ahead to better prepare for what might be coming. Today, global climate change and its local implications have institutionalized the search for a window to the climate future. Many scientists believe that society can improve its relationship with atmospheric processes by producing better forecasts on a variety of time scales. The truth is that improved probability-based technical forecasts for future states of the atmosphere supply only part of the answer. The other part of the answer will come from a better understanding of how climate interacts with human activities.

While scientists, supported by their governments and research establishments, have been preoccupied with trying to understand how the climate system works, they have devoted considerably less time to identifying ways that forecasts can be used to address specific societal needs. During the past three decades or so, several social scientists have undertaken research on how to use weather and climate forecasts to the benefit of society. Forecast value studies undertaken in the 1990s have built on the earlier works of "impacts pioneers": Jack Thompson, James McQuigg, and Stanley Changnon, among others.

Today, we have one community of researchers that focuses on the scientific aspects of forecasts and another community that focuses primarily on the needs of forecast users. There are good reasons to link these if governments want to maximize both the value of forecasts for societal use and the effective application of the science of forecasting to societal needs.

A key reason that some governments spend so much money to develop reliable forecast mechanisms is that they want an early warning capability to identify climate-related problems. Its purpose is to reduce potential adverse impacts of climate-related episodes on the country's physical infrastructure and its citizens. In a developing society, forecasters focus on food production. Their purpose is not necessarily to reduce the number of people in poverty in the country, but to prevent severe food shortages and famine. Climate-related forecasts alone cannot resolve many of the chronic socioeconomic problems faced by present generations (e.g., poverty, hunger, population, and AIDS crises). Forecasts are a highly visible example of how climate-related information can be used to benefit the well-being of all members of society.

SIX

HOW WE KNOW WHAT WE KNOW

This chapter presents some examples of attempts to identify which socioeconomic or environmental impacts can legitimately be associated with or blamed on climate anomalies or weather extremes. Any given impact is usually followed by the claim that "Mother Nature did it" and later by a claim that the impacts resulted from poor decision making. However, attributing an impact to an atmospheric or societal cause is not an easy task. Many scientists have attempted to develop methods that improve the way they make those attributions. Although none is perfect, there is a need to sort out what part of causes of any particular impact can be associated with nature and what part with society. This is a very important task because policies to improve the societal interactions with atmospheric processes depend on the results of attempts to link cause and effect.

There is no single best research approach to identifying the impacts on society of either normal or anomalous climate. In fact, there are many useful methods. Some are highly quantitative, and others are qualitative. Regardless of the research tools used for climate impact assessment, they are geared for the most part toward improving society's ability to correctly attribute specific adverse impacts on society and the environment to climate and climate-related anomalies. In the past two decades or so, many assessment methods have also been developed to help identify ways for societies to improve how they cope with potentially harmful impacts of climate anomalies and weather extremes or to capitalize on climate's resource aspects.

Quantitative and Qualitative Methods to Assess Impacts

Some researchers rely on computer models to generate, for example, scenarios of potential impacts of climate variability, change, or extremes on ecosystems or on socioeconomic sectors. Others use a historical or an analogue approach by looking at how societies have responded to climate-related impacts in the past. They seek to identify societal coping mechanisms that might have application under present-day circumstances.

Climate impacts can be monitored on the ground, from the air, and from space—and even from tabletops in libraries and offices. Each of these data platforms provides unique and reinforcing information. Satellite imagery can be used to detect and monitor changes in land cover over time, as has been demonstrated, for example, in Brazil's Amazon rainforest or in Central Asia's Aral Sea basin. Local histories of vegetation and land-use changes, based on anecdotal, qualitative, and quantitative data, can be used to uncover the earlier history of land transformation in these and other specific locations.

Reviews and analyses of historical events, which have recently become popular for carrying out, for example, El Niño, drought, and flood impact studies, are extremely useful in weather- and climate-related impact assessments. Historian Fischer (1996) suggested:

> The study of history can never tell us with certainty what will happen next, but it gives us the benefit of much hard-won experience in the past. It also helps us to know our intentions for the future (p. 236).

Anecdotal information can sometimes be as important as scientific information and hard data. Stories about how the land and its vegetative cover have changed over time provide rich accounts of subtle forcing factors and changes that might not have been captured by other methods. Novelist Chaim Potok (2001) wrote that "without stories there is nothing. The past is erased without stories" (p. 74). (On the use of analogies and qualitative information for impact assessment purposes, see Glantz, 1988 and Jamieson, 1988.)

Several attempts have been made in the past two decades to compile methods for climate and climate change-related impact assessment (e.g., Kates et al., 1985; Riebsame, 1988; Feenstra et al., 1998; see also unfccc.int/program/mis/meth/view.html). Organizations that fund such efforts believe that a compendium of assessment methods can be kept on a bookshelf as a reference, much like a cookbook, awaiting the right time for use by researchers in various disciplines, socioeconomic sectors, and countries. Because most impacts are problem- or location-specific, either of those factors, or both, often determine the methods for analysis. Given the increasingly technological nature of and bias toward new research methods, we must be careful that the methods we use do not limit how we research climate-related environmental or social problems. Method selection must be determined by the problem to be addressed as much as by the new methods being developed to address such a problem. The latest approach to climate studies has taken the form of integrated assessment.

Integrated Assessment

Integrated assessment (IA) is a methodological concept that has many definitions. Rotmans and Van Asselt (1996) define it as "an attempt to bring together various knowledge domains in order to synthesize these pieces of information into insights that cannot be derived from a single disciplinary analysis."

Integrated assessment of global environmental issues is a new field that has been emerging over the past few decades. In recent years IA has received considerable attention in the climate change research community (*Climatic Change*, 1996; Rotmans and van Asselt, 2001). It is based on the assumption that causes, effects, and responses to environmental changes (in this case, global warming) cannot be best identified through disciplinary or multidisciplinary assessments. For example, the study of a specific meteorological drought (say 75% of average rainfall) will likely evoke different policy responses, depending on the disciplinary perspective used to assess it. A meteorologist might call for cloud seeding in an attempt to increase precipitation; water resource experts see the solution in terms of more reservoirs; economists may suggest that farmers plant crops that are more tolerant of moisture stress; politicians may favor responses that are highly visible to voters but are not necessarily very effective as drought response. IA's objective is to bring all disciplinary perspectives (referred to as a unifying approach) to bear on complex environmental issues. By doing so, the whole can prove to be greater than the sum of its disciplinary parts.

Integrated assessment is a goal toward which researchers involved in environmental assessments, including climate assessments, hope to move. However, it is a nascent sub-field striving to work past its constraints and to capitalize on its benefits. Some researchers have expressed caution about an over-reliance on IA as a new approach just because it has been presented as *the* approach to capture all aspects of an environmental issue (Risbey et al., 1996). Yet, its value as a tool to inform policymakers has yet to be proven. In the meantime, many researchers continue to rely on interdisciplinary, multidisciplinary, and disciplinary research methods, each of which (like IA) has its own strengths, weaknesses, opportunities, and constraints.

One of the major criticisms of IA has been its dependence on quantitative knowledge and approaches. On this point, Rotmans and van Asselt (1996) suggested that "only by including the social scientific perspective in IA can its scope be broadened and deepened, and the quality of current IA methods improved" (p. 333). They also observed that "the process of decision-making, the form of actors values, preferences and choices cannot (yet) be captured by formal models" (p. 330).

Depending on one's level of confidence in the results of integrated assessments or in the quantitative modeling exercises on which many IA studies are based, one can use IA results either in decision- and policymaking processes to guide policy decisions or as an imprecise tool that can help to identify neglected aspects or new insights relating to the specific issue being assessed.

Foreseeability

Foreseeability is a concept that has been used by the legal profession for more than a century. According to one dictionary (Gifis, 1991), it is used in various areas of the law to limit the *liability* of a party for the consequences of his acts to consequences that are within the scope of a *foreseeable risk*, i.e., risks whose consequences a person of ordinary prudence would reasonably expect might occur. In tort law, in most cases, a party's actions may be deemed *negligent* only where the injurious consequences of those actions were foreseeable.

The definition of foreseeability includes two aspects. One aspect refers to the concept, which means to see beforehand or to foreknow, and the other aspect refers to the application of the concept to determine responsibility for damage caused by the consequences of a foreseeable (and therefore avoidable) event. With respect to the first aspect, the concept of foreseeability can be extremely useful for identifying possible climate impacts on human activities and on the environment. It can also be used to identify possible impacts of human activities on the climate system. It differs from "predictability" or "forecasting" because it neither depends on nor implies a quantitative description of the probability of harm. It suggests, for example, that a reasonable person can conclude that certain climate or weather anomalies could likely have certain knowable adverse impacts on human activities and environmental quality, leading to soil erosion, deforestation, salinization, bush and forest fires, or destructive flooding.

The second aspect of the concept, responsibility, relates to tort law in that it is used to identify parties responsible for causing harm. Terms used in the legal definition, such as "ordinary prudence" and "person of reasonable intelligence," are subject to interpretation and may prove to be situation-specific. So each time forseeability is raised as a legal issue, the context in which a situation was deemed foreseeable and a determination of prudence and reasonable intelligence must be identified if a legal precedence does not already exist. However, these two aspects are detachable if one is seeking only to identify, through the concept of foreseeability, a potential future harm to people or to the environment. Laying blame and seeking compensation for the harm that might have been foreseeable to a reasonably prudent person is another matter.

For example, "the foreseeability element of proximate cause is established by proof that the actor or person of reasonable intelligence and prudence, should reasonably have anticipated danger to others created by his or her negligent act" (Gifis, 1991, p. 195). The negligence is that of the authority who failed to have foreseen adverse impacts. Recall that inaction is also a type of action. "Foreseeability encompasses not only that which the defendant foresaw, but that which the defendant ought to have foreseen" (Gifis, 1991, p. 196).

Thus, it is foreseeable that irrigating arid and semiarid lands can lead to a range of adverse impacts, if not properly planned and carried out. Impacts will appear first in the environment (e.g., waterlogging, salinization) and then in settlements (e.g., reduced crop yields, reduced water quality, adverse health effects of pollutants). These kinds of changes can eventually lead to the abandonment of the land and settlements and the desertification of the landscape if no actions are taken to deal with them.

Another idea linked to foreseeability is that of "present futures." This is the use of recent climate-related impacts as analogies, that is, a search for examples of possible future changes and impacts in one location that already exist in other similar locations. Present futures provide a reality check with regard to realistic projections of societal and environmental changes that might result from climate anomalies. They would be case-based scenarios—situations that have already taken place somewhere else or in the recent past. Reviewing the responses of decision making authorities to the forecasts or impacts of adverse climate- or weather-related events can help researchers and political leaders identify good climate-related decisions and bad ones. In the earlier situation, the adverse impacts of climate-society interactions might not have been foreseen, but that would not be the case in the latter situation. This could be viewed as a *foreseeability analysis*.

Attribution: What Impact to Blame on Which Anomaly

Whenever an extreme weather event or climate anomaly occurs, people throughout society, from citizens to national policy makers, tend to blame many existing problems on it. That tendency is probably a human response, a desire to find something or someone to blame. We seem to need to know the sources of the problems we face. It appears to be less important to identify a true cause than to feel comfortable with a cause that is plausible. The case is then closed until a more plausible explanation emerges.

Although a single factor might dominate as a cause, it is usually a combination of factors that makes a stressed situation worse. For example, many people still tend to blame automatically a specific drought for an ensuing famine. Although a drought can lead to serious food production problems and to food shortages, it does not have to lead to widespread migration, refugee camps, and starvation deaths. As noted earlier, West African governments decided to export food products in the early 1970s, knowing full well that their citizens were in the midst of extreme drought-related food shortages. It was also the governments' decision to continue to grow cash crops for export on the country's better watered lands, instead of turning that land over to food crop production to meet national food needs. The decisions to continue to export cash crops contributed to a truly regional famine (Lofchie, 1975).

Another example of the attribution problem is the 1998 floods in China's Yangtze River basin, one of the country's worst floods in the twentieth century. They adversely affected one-quarter of the country's billion-plus population. At first, government officials blamed the floods on a slow-moving storm system that dumped a considerable amount of rainfall in the middle reaches of the river system. Later, researchers identified widespread deforestation of the slopes in various locations in the upper reaches of the river basin as a major contributing factor to the devastating floods. To its credit, the Chinese government later admitted that deforestation activities over a forty-year period had increased the basin's susceptibility to rapid rainfall runoff, as the soils in the deforested areas were no longer capable of holding much moisture. Heavy rainfall also played a role in the flooding. However, yet another factor, the heavy snowfall and the cold winter during the 1997–1998 El Niño, was neglected as a possible contributor to the floods. The combination of these prevailing conditions provided more water to the river system later in the year than normal. If it can be shown that El Niño did have a major influence on the amount of snowfall on the Tibetan Plateau, and therefore on the subsequent heavy runoff, then early warning of El Niño could prove to be very useful in managing the flow of the Yangtze in both El Niño and La Niña years. One of the key lessons here is that it is important to get the attribution of cause to effect correct before making new policies.

Most recently, the drought-famine-war-religious fundamentalism nexus in Afghanistan from the late 1990s to 2001 reinforced the view that, although drought by itself can do considerable damage to a country's rates of morbidity, mortality, and migration, political and religious decisions by the ruling Taliban authorities had led that country down a road to famine.

Today, many authors seek to identify a climate factor to improve on present-day interpretation and understanding of key historical events. For example, the recently discovered phenomenon of El Niño seems to be getting a lot of attention in retrospective studies that seek to identify its role in turning points in history. Writers have been providing alternative climate-related theories about why certain major historical events occurred. For example, one researcher suggested that the French Revolution in the late 1700s had been influenced by an intense El Niño event. Another researcher linked famines in colonial India, China, and Brazil to El Niño-spawned droughts (Davis, 2001). The problem with this newfound interest in climate anomalies, especially El Niño, and their influences on historical events is that it has reinforced searches for a single, dominant climate influence on important socioeconomic or political situations of the past, where perhaps none existed.

Although climate is an important aspect to consider, in some cases it was not so important to the historical or future course of events, whereas at other times it may have been the determining factor. Natural and social scientists must study the problems associated with making reliable attributions of turning points in history to weather or climate anomalies (see box 6.1).

BOX 6.1.

El Niño and the Problem of Attribution

One of the earliest attempts to link El Niño to change in the course of history was made by American geologist Sears (1895). Pizarro and his conquistadors invaded northern Peru at Tumbez from 1529 to 1531. Pizarro attacked the Inca Empire, bringing about its collapse. Sears attributed this success in part to an El Niño (referred to by the Inca as "the septennial rains"), based on anecdotal accounts of the periodic availability of vegetation in the normally desert environment of northern Peru and on accounts of malaria in the Chiura Valley. Such environmental changes usually accompany El Niño's heavy rains. Hence, in Sears' view, Pizarro's conquest of the Inca was made possible by an El Niño episode.

Scientists now use sophisticated computer models and their current understanding of El Niño to forecast the likely onset of the event. However, their model-based projections missed the mark as recently as February 1997. By then, the early onset of an El Niño event was already beginning to be observed by satellites, buoys, and other methods of detection. So even with the best equipment ever, the research community was unable to forecast this El Niño. However, this El Niño event was the best *observed* ever.

As for the impacts of climate anomalies that occurred during the recent 1997–1998 El Niño, researchers five years later are still trying to sort out what they can legitimately attribute to El Niño from what can be blamed on human activities. Researchers have trouble identifying the true causes of many of the climate-related disasters around the globe. Why, then, do researchers believe that they can do it for historical El Niño events that occurred more than a hundred years ago? Why is our hindsight for such historical inter-actions seemingly 20/20, when our "now-sight" is nowhere near that level? People can speculate, hypothesize and investigate earlier events but to attribute *with confidence* cause (e.g., El Niño) to effect (e.g., famine) remains very risky.

Most people believe that more information about the future is of great value. Scientists reinforce this thinking. Yet, existing informa-tion about the climate system is not automatically valuable. For the most part, the value added will come from analysis and a judicious use of that information at the right time and in the right context.

Psychological Aspects of Climate and Weather

Psychology is the scientific study of the behavior of humans and animals. Psychologists use scientific methods in an attempt to understand and predict behavior, to develop procedures for changing behavior, and to evaluate treatment strategies. (NMU, 2000)

The website from which the above definition was taken also noted the difference between social and cognitive psychology: Social "is concerned with the effects of social situations on human behavior," and cognitive is concerned with "memory, thought, problem solving, and psychological aspects of learning" (NMU, 2000). Climate and climate-related perceptions have been a focus of research for a couple of decades (Whyte, 1985). Thus, it is easy to show why psychology has a key role to play in a wide range of climate and climate-related issues.

In fact, psychologists have already been involved in research on environmental issues in general and on climate and climate-related issues specifically. Some researchers have focused on perceptions of extreme events, such as droughts, hurricanes, and floods (White, 1974), on perceptions of the credibility and value of forecasts (e.g., Stewart, 1997), and on perceptions of weather and climate risks, extremes, and impacts (e.g., Saarinen, 1966; CSTPR, 2002). Others have focused on public perceptions of atmospheric issues, such as global warming and stratospheric ozone depletion (Fischhoff, 1994). Still others have focused on "ecopsychology," a field that, according to Roszak et al. (1995), represents a new generation of psychotherapists who are "seeking ways in which professional psychology can play a role in the environmental crisis of our time" (p. 3).

A well-noted example of how perceptions of climate can influence human and governmental behavior appeared at the outset of the twentieth century. During this time, climate conditions in a given region were viewed as a boundary constraint on economic and social development. Ellsworth Huntington (1915) articulated this perception; he believed that people in hot tropical climates are less productive and energetic than those living in the midlatitudes. This view was attacked by many writers as racist but has since been repeated, and not just by those who live in the midlatitudes (e.g., Bandopadhyaya, 1983). Referred to as climate determinism, such perceptions have influenced governmental, institutional, and individual behavior and policies of the wealthier midlatitude countries toward tropical developing nations. Even if one's perception of reality is erroneous, actions taken based on those misperceptions will have real consequences.

Another useful example is a study produced in the 1960s, when geographer Thomas Saarinen completed his study of the perceptions of drought held by farmers in the American Great Plains. He found that young farmers who had not experienced severe droughts had different attitudes toward the way they managed their farms and different views about the level of drought risk compared to the older members of the farming community who had witnessed the Dust Bowl of the mid-1930s. Even if the young farmers were aware of the adverse impacts of the Dust Bowl days, they may have come to believe that new technologies and land-use practices would help them cope effectively with future drought situations.

Interestingly, in the early 1970s, when droughts and food shortages were occurring in many locations around the world except in the U.S. grain belt, American farmers began to cut down the shelterbelts (trees and hedges) that had been planted in the 1930s to break up the flow of desiccating winds. They did so to take short-term advantage of the high grain prices in the marketplace, perhaps perceiving that new technologies and land-use practices would prevent the possibility of a return to the Dust Bowl days. What they failed to realize was that their actions to cash in on short-term benefits had increased the risk of dust storms when future drought conditions returned, as they surely would. This is also an example of a human tendency to reduce the value of historical lessons to present-day actions.

Writing about psychological aspects of global climate change, Rachlinski (2000) noted the following:

> Judge Learned Hand asserted that a reasonable person takes any precaution that is less burdensome than the probability that some harm will occur multiplied by the magnitude of the harm. Presumably, a reasonable society

does the same. A reasonable society should be willing to undertake fairly significant precautions to avoid cata- strophic events. . . . Over the past few decades, however, social and cognitive psychologists studying human judg- ment and choice have learned that reasonable people sometimes fail to make reasonable choices. Cognitive limitations on human judgment and choice sometimes lead people to make decisions that produce unwanted outcomes.

Rachlinski went on to note that "one can scarcely find a problem faced by contemporary society that better fits the definition of a social trap than global climate change" (p. 1).

Figure 6.1 summarizes the results of a survey of university students who were asked to associate verbal statements with their perceptions of quantitative probability statements.

Although probabilities are discussed frequently among researchers, the public tends to rely on qualitative expressions of probabilities or likelihood of occurrence. Because an individual's perceptions are formed by a subset within a wide range of factors, there is no precise equivalence of a specific numerical expression of probability to a specific verbal statement. Also, words in one language simply do not convey the some meaning in other languages. For example, at a recent workshop in South- east Asia, some participants equated the occurrence of a "highly likely" extreme climate event with a 30–40 percent probability.

Studies about human perceptions of climate-society interac- tions have identified some perception-based behavioral responses to global warming concerns: for example, people who are risk- averse will likely respond to information on global warming in ways that are quite different from those who are risk takers. People tend to view uncertainty in scientific information about climate change in different, often opposing, ways. If the level of uncertainty is considered to be high or the rate of change rela- tively slow, then there is little reason to act quickly. If the level of certainty is perceived to be high or the rate of change relatively rapid, there is a heightened tendency to act more quickly.

Scientists have used surveys to identify public perceptions about different aspects of the climate change issue. As an example,

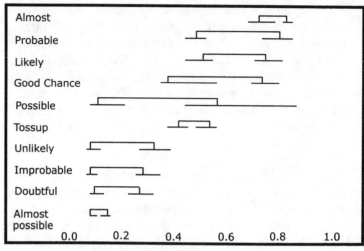

Probability

Figure 6.1. Upper and lower probability limits for probability expressions. Upper bars show mean lower and upper probability limits provided by students for each verbal expression of perceived likelihood of occurrence. Lower left- and right-hand bars show the interquartile ranges provided for the lower and upper limits, respectively. (Source: Fischhoff, 1994)

scenarios were presented to respondents who were asked about the amount of money they would be willing to pay to avoid certain climate change-induced environmental changes. The following question was posed: "Would you be willing to pay x dollars more a year for the things you normally buy in order to prevent this climate situation from occurring?" The author of the survey identified the biggest surprise: it took big climate changes to budge people. As was pointed out, public perceptions are reality to individuals. For example, one survey participant noted that after the Valdez oil spill, the quality of the salmon was fine, but people wouldn't buy or eat it, because of the perception that its quality was low. In economic terms, the *perceived* quality was what mattered (Berk, 1994).

Again, with regard to the global warming issue, a small but vocal group of physical scientists do not believe the scientific information used to support the view that human activities can heat up the global climate (e.g., Huber, 2000). Their perceptions, however, are based on selective attention, choosing to pay attention only to scientific information that supports their beliefs. One global warming critic suggested that sea level *will not* rise with global warming. Although this view goes against the general consensus of the scientific community, it bears mentioning.

According to Fred Singer (no date),

> recent observations and new analyses of existing data suggest an opposite outcome: *A climate warming should slow down sea level rise not accelerate it.* To understand this counter-intuitive result, one must first get rid of false leads—just like in a detective story. The misleading argument here is the oft-quoted statement that the climate warmed by 1°F (0.6°C) in the last 100 years and that sea level rose by 18 cm. Both parts of the statement are true; but the second part does not necessarily follow from the first.

Singer's analysis flies directly in the face of virtually all other studies. Nevertheless, this statement, serves to underscore the reality that diametrically opposed perceptions of global warming impacts exist.

Although Singer has challenged the science of global warming, psychologists have identified reasons why the public has shown a reluctance to act on existing reliable scientific evidence that supports a global warming hypothesis. Rachlinski (2000) has suggested,

> Several psychological phenomena of judgment . . . support a more pessimistic perception on humanity's ability to respond effectively to the prospects of global climate change: the lack of scientific consensus and the reluctance to cutback on their economic status quo, e.g., to change their status quo for the worse. Psychologists have referred to this as "loss aversion." Studies suggest that individuals are more likely to sacrifice to protect their environment than they are to give up something to bring the environment back to a level that they would

like to see. . . . In other words, people are more willing to gamble to avoid a loss than to obtain a benefit.

Climate Ignorance Versus Climate "Ignore-ance"

Many people are unaware of various aspects of the climate system or of the ways that atmospheric processes interact with the environment and with human activities. Perhaps the adage that ignorance is bliss is appropriate; the general public around the globe is focused on the day-to-day problems of going to work, feeding families, and making ends meet. I am not at all troubled by such a lack of climate awareness on the part of the general public, as long as others in society—what columnist Walter Lippmann (1955) once referred to as the "attentive public"—take on the responsibility of monitoring climate issues on behalf of the inattentive public. The attentive public comprises a relatively small percentage of any society whose attention and concern are focused on major issues of importance to the country. Thus, the general public is, in essence, being represented by the attentive public, which has the time, energy, interest, and resources to engage in debates on contentious climate and climate-related issues.

What is troublesome, however, is "ignore-ance." This fabricated word refers to people who are aware of the potential for adverse climate-society-environment interactions but, for whatever reason, political gain or economic profit, choose to do nothing about it. For example, when it comes to wetlands, there has been a constant tug-of-war between those who favor the use of wetlands for construction purposes and those who seek to preserve them as ecological reserves for enjoyment of future generations.

More than a decade ago, then-U.S. Vice President Dan Quayle attempted, with the stroke of a pen, to redefine the term "wetlands." Instead of defining it as a location with nine consecutive days of standing water, he wanted to increase the number of days to sixteen. This would have reduced by half the total area of officially designated wetlands in the United States, thereby opening up the other half to land developers. His attempt to redefine the accepted definition of "wetland" was rejected by the attentive public. This example shows that even a seemingly innocent redefinition of a concept can have devious political underpinnings with major environmental implications.

The climate change issue also presents examples of ignore-ance. For example, whether the global climate is changing is really no longer an issue (average global temperature has increased in the twentieth century by 0.7°C [1.3°F]), but disagreement still exists about what portion of that change results from increasing societal emissions of greenhouse gases. One cannot deny that a dozen of the hottest years on record have occurred within the past two decades, with the more recent years setting new records. We do know that the burning of fossil fuels and tropical deforestation increase the atmospheric concentration of CO_2. Nevertheless, many of those in the energy sector or those engaged in the wanton destruction of tropical forests, although aware of the science of global warming, choose to ignore it. Therefore, they do not alter their greenhouse gas-emitting practices. They, too, along with many government leaders, suffer from ignore-ance.

Thus, it is very important for those already involved in climate affairs to help broaden the climate-related knowledge base of the attentive public and, perhaps more importantly, to generate awareness of and interest among the inattentive public about the various ways that climate-related issues can affect their families. Emphasizing relevancy can reduce igno-rance among the general public, while also reducing ignorance for decision makers by making it more difficult to purposely ignore the many ways that climate variability, change, and extremes influence human activities, human settlements, and the environment.

Does Climate History Have a Future?

Many methods are available to identify how the climate system operates at present, how it worked in the past, and how it might operate in the future. They range along a continuum from highly technical and sophisticated satellite imagery to highly uncertain but potentially useful information derived from surveys, interviews, and anecdotal sources. In the middle of this range of methods, one can find approaches to determine prehistoric climate regimes. Many scientists today are committed to reconstructing climate conditions of the past in an attempt to better understand possible climate conditions of the future. Data come from a variety of disciplines, and provide a piece of the climate system puzzle. For example, paleo-proxy data include the following: measurements of the chemical composition of coral, which changes as a function of water temperature; fossilized pollen taken from soils and beds of ponds, lakes, and the ocean, which provides insights into the kinds of plants that grew in the region under various climate conditions; ice cores containing various trace gases taken from glaciers and polar ice caps; and tree-ring analyses, in which the number and width of rings reflect annual climate conditions over time (wet and dry periods).

Historical records of activities such as wine harvest dates or palm oil production, along with ecological information such as the annual dates over centuries of the first appearance of cherry blossoms, have also helped scientists to reconstruct past climate regimes. Personal records such as diaries, travelogues, and other qualitative accounts have also provided useful insights about climate conditions that prevailed in the past.

In some locations, researchers have been collecting quantitative information about rainfall and temperature for well over a century or two. Some cities in Europe (e.g., London, Budapest) have recorded this information for more than 200 years. For the most part, reliable information has been systematically collected for many locations in the tropics only for decades. Nevertheless, many interesting and important writings about climate and climate-society interactions in developing areas have been completed in the past two centuries. These early works provide reliable quantitative and qualitative information about human interactions with the land and the climate. We now know with considerable reliability about extreme droughts and floods in locations such as India and China over the past thousand years. Ethiopia and Egypt have centuries of reliable historical information about food shortages and Nile floods, respectively.

Several authoritative climate histories have been written that provide useful, detailed analyses of climate-society interactions. For example, Brooks (1926) prepared *Climate through the Ages*, which provides insights on climate up to the early part of the twentieth century. It also provides many references to writings that Brooks used in the preparation of his book. Decades later, Hubert Lamb (1982) prepared several climate histories. His book, *Climate, History, and the Modern World*, provides valuable accounts of the climate record, processes, and impacts over the centuries. Ladurie (1971) judiciously used proxy indicators to reconstruct climate history and the interplay between climate processes and human activities such as agricultural production over the past millennium. Bryson and Murray (1977) presented their evidence for the influence of climate variability and change on the fate of some civilizations.

In 1978, Canadian climatologist F. Kenneth Hare, author of *The Restless Atmosphere*, made the following observation:

> Climate has been one of the many historic influences that historians have failed to take seriously. Or in some cases they have taken it too seriously, while lacking the factual basis to back up their theories. . . . In the last few years, a small but precocious breed of climatic historians has transformed our knowledge of the past thousand years. . . . Nothing could be more timely, for climate is once again at the heart of the matter (p. ix).

Today, that situation has changed. Historians are now among the growing community of researchers from many disciplines involved in various aspects of climate-related research. Yet, something is still lacking in climate research. It relates to the research findings of previous generations that the present generation of researchers chooses to pass on, filtered, to the next generation.

Today's students of climate issues most likely are not being made aware of research activities of a generation ago, such as the Climate Impact Assessment Program (1970–1975), a policy-driven research program that produced a blockbuster study, perhaps the first of its kind. The U.S. Department of Transportation evaluated the impacts on stratospheric ozone of the chemical emissions of a large fleet of high-flying supersonic transports (SSTs). Today's students are most likely unaware of the controversy that immediately emerged between the United States and the European atmospheric research communities over the science surrounding ozone depletion and the SSTs (Martin, 1979).

Are young researchers today being made aware of the concern of atmospheric scientists just over two decades ago about the possibility of a global cooling of the earth's atmosphere? In the early 1970s, following a three-decade period of below-average global temperatures, scientists raised the possibility of an impending climate cooling. Some referred to it as the beginning of a return to an ice age. Various researchers cited convincing but circumstantial evidence in support of the cooling hypothesis. Evidence included the following:

- between 1940 and 1970, the British growing season was shortened by two weeks,
- fish populations normally caught off the northern coast of Iceland were being found off the country's southern coast,
- hay production in Iceland was sharply reduced,
- armadillos, which had migrated as far north as Kansas, began to retreat toward the south, and
- an increasing amount of Arctic sea ice was found in normally ice-free shipping lanes of the North Atlantic.

The titles given to several books and articles in the mid-1970s suggest the level of concern about a possible climate cooling: *The Weather Conspiracy: The Coming Ice Age* (The Impact Team, 1977), *The Cooling* (Ponte, 1976), *Forecasts, Famines and Freezes* (Gribbin, 1976), "When the Sahel freezes over" (Ponte, 1977), and "Ominous change in the weather" (Alexander, 1974), among others.

The circumstantial evidence was convincing enough at the time to cause the U.S. Central Intelligence Agency to prepare strategic reports on the impacts of a cooling on the energy and agriculture sectors of its archenemy, the Soviet Union (U.S. CIA, 1974). The U.S. Congress also held hearings on the prospects of global cooling and its implications (U.S. House of Representatives, 1976). A keynote speech at the First World Climate Conference held in 1979 focused on the economic aspects of climate cooling (D'Arge, 1979).

The global cooling hypothesis gave way in the mid- to late-1970s to a relatively small but growing concern among scientists about global warming. To many older researchers, the global cooling issue is a distant, if not unwanted, memory. Some of the researchers who were supporting a global cooling hypothesis have since become global warming proponents.

The IGY (1957–1958), the National Hail Research Experiment, (1970–1975), Project Stormfury, Global Atmospheric Research Program (GARP), GARP Atlantic Tropical Experiment (GATE), Monsoon Experiment (MONEX), and West African Monsoon Experiment (WAMEX), among others, are climate-related research projects of the previous generation. Each produced useful scientific information, much of which should be reviewed by new generations of researchers for ideas meriting further study.

Researchers involved in climate and climate-related impact studies need to look more closely at such earlier studies. They provide a wealth of references to even earlier works on which the studies were based. They also provide the reader with hypotheses that may have been overlooked in earlier times because of the lack of present-day understanding and insights about climate-society-environment interactions. As one generation of researchers passes the proverbial baton to the following generation, the new generation tends not to look very far back in history. So a new researcher seeking to identify the value of climate or weather forecasts, for example, might not believe that much useful work on the topic had been done before, say, 1990. Perhaps students are led to believe that the most important aspects of the work of their predecessors are well summarized in recent publications. Yet, information we are given today is extracted from a foundation of research findings and has been filtered by preceding generations of educators and researchers. This suggests that many insights may have been unwittingly passed over. Several examples show that today's students of the climate system are deprived of, and are losing touch with, important details embedded in past research write-ups.

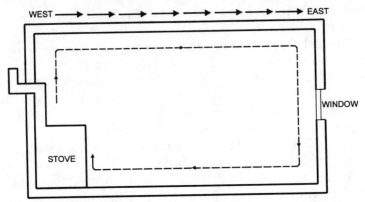

Figure 6.2. Simplified graphic of global atmospheric circulation from west to east across the tropical Pacific depicted as circulation of air in a closed room. "The heating of the atmosphere drives a cloud system (called a convective cell). Low-level air flows across at the base (from right to left), is heated within the cell, and rises and flows out (from left to right) at high level." (Source: Gill and Rasmusson, 1983 p. 231; Ghil and Childress, 1987)

Example #1: Researchers used figure 6.2 in a 1987 article to depict in a simple way the circulation of the atmosphere in the tropics, now called the Walker Circulation.

At first I thought how clever it was to include such a simple diagram of a complicated atmospheric process. However, I accidentally found a similar graphic (figure 6.3) in a 1904 geography book to explain atmospheric circulation; this was published several decades before that atmospheric phenomenon was named after Sir Gilbert Walker.

Example #2: Figure 6.4 appeared on the AccuWeather website in 2001. The graphic depicts the frequency of tornadoes in the United States, based on post-World War II data.

While reviewing an 1888 geography book, I came across figure 6.5, a representation of the frequency of tornadoes in the United States to that time. Considering that weather scientists a century ago depended on tornado sightings by individuals, and given the relatively sparse population west of the Mississippi River at the time, the graphic appears to be quite similar to the one produced in 2001. However, the earlier

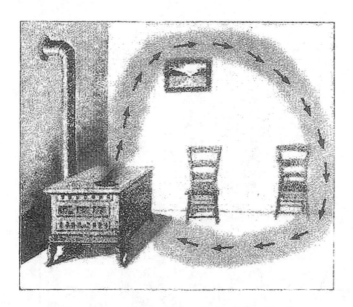

Figure 6.3. An artist's rendition of atmospheric circulation in a closed room as idealized circulation of air across the Pacific at the equator. (Source: Tarr and McMurry, 1904)

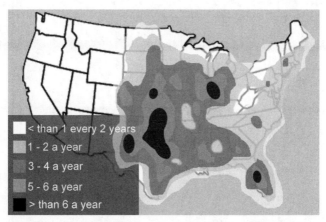

Figure 6.4. Tornado frequency in the United States, 1950–2000. (Source: AccuWeather, 2001)

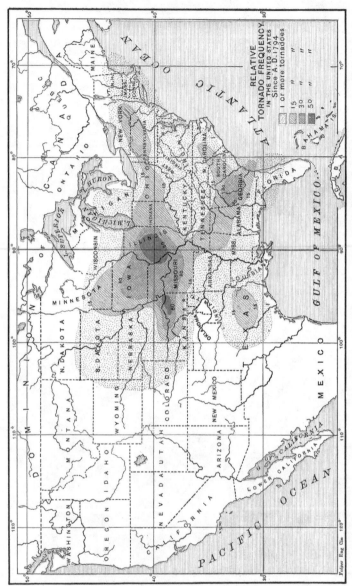

Figure 6.5. Tornado frequency in the United States. Note that the information used to construct tornado frequency dates back as far as 1794. (Source: Hinman, 1888)

one was based on whatever information was available at the time; interestingly, the data extended back to as early as 1794! To be sure, today, researchers would not consider using such old and anecdotal information in their research activities.

Example #3: According to Bill Gray, Colorado State University hurricane expert, Gordon Dunn was the first to express awareness in 1940 about the likely genesis of hurricanes off the African coast. Many of the hurricanes in the tropical Atlantic are now known to have their origins off the West African coast. A pull-out chart in a report on Atlantic hurricanes published by the American Association for the Advancement of Science in 1855 (Redfield, 1855) showed the tracks of numerous hurricanes that had occurred between the end of the 1700s and 1854. Although obviously not a complete record of tropical storms for this period, figure 6.6 does show that many of them had their origins off the West African coast.

There are many other examples of recent climate-related events that could be compared to earlier ones: prolonged and devastating droughts in South Africa in the 1910s and again in the 1980s and 1990s; Mississippi River floods in 1927 and 1993; monsoon failures in India in 1877–1878 and in 2002; Yangtze River floods in the first half of the 1900s and in 1998; and so forth. Historical accounts of climate anomalies (their causes as well as their impacts, ecological or societal, and responses to them) can provide valuable information even today for societies that must cope with such anomalies. These accounts can be used to identify what institutions have learned or failed to learn over time about climate-environment-society interactions in their region.

Thus, it is not only useful but also necessary for students of climate to review for themselves the original writings of their scientific predecessors in the physical, biological, and social sciences. Not to do so makes students dependent on the brief summaries of the earlier works that have been prepared by other scholars who pick and choose what they think should be passed on to future generations. Despite the fact that research methods have improved and mountains of information have been compiled, there are likely to be benefits to those who choose to make sure that historical information does have a future in research efforts to come.

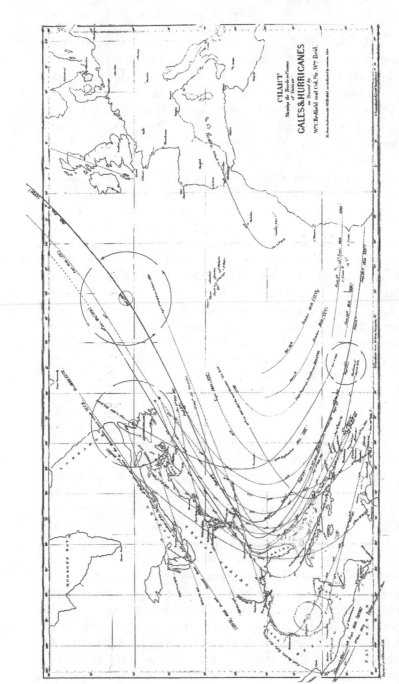

Figure 6.6. Tracks of tropical Atlantic and Gulf of Mexico hurricanes before 1854. (Source: Redfield, 1855)

SEVEN

CONCLUSION

Climate Dimensions of Social Science

A few years after the physical and biological science research community developed a science-focused global change program, the International Geosphere-Biosphere Programme (IGBP) in the mid-1980s, they created a program for the "human dimensions of global change." This was done in response to concerns expressed in various government and academic quarters that the impacts of human activities on Planet Earth and, more specifically, on the structures and functions of its different ecosystems, were not being taken into account in global change research. Thus, a human dimensions program was spawned more by a political necessity to humanize a scientific research agenda than by a belief that the social science contributions were needed before scientific issues had been successfully addressed.

The phrase "human dimensions" reinforces the view that the inclusion of human activities in the IGBP research agenda was an afterthought of geophysical and biological scientists. The societal aspects were considered to be supportive of physical and biological global change research. However, funding today by many governments for scientific programs and projects increasingly requires the involvement of the social sciences.

A recent IGBP report asserted that the concept of global change was meant to encompass much more than just climate change (GECP, 2001). Swedish researcher Svedin (2001) commented on the change in views about the inclusion of social scientists in global environmental change research. He suggested that, "at an early stage, this component of knowledge was not seldom seen as an end-of-the-line assessment of impacts, for example, economic impacts, related to study outcomes based on natural science traditions. This is not so any more. Human dimensions issues come in much earlier nowadays and in much more integrated ways" (p. 10).

Until recently, the human component of the IGBP had been viewed as an appendage to a multibillion-dollar scientific research agenda. However, as this book presents throughout, human settlements and their activities have become factors that can force changes in the global climate system. This raises an interesting question: Is it appropriate and useful to examine the climate dimension of human activities? As an experiment, let's discuss the "climate dimension" of social science.

Every society is confronted on a continual basis by a wide range of issues with which it must contend: fiscal, cultural, legal, political, ethical, economic, health, conflict, and environmental issues. Each issue is likely to have at least one important climate dimension, its physical climate setting.

In the past, societies that developed in, for example, arid environments with low, erratic rainfall and a resultant desertlike landscape, including a fragile vegetative cover, had to devise ways to live with those conditions. Irrigation was one key response, bringing water by gravity flow at first from mountainous areas to parched soils (Wittfogel, 1974). This enabled a society to flourish in an otherwise harsh environment. People in mountainous environments, such as the high Andes, devised ways to grow food crops and raise livestock under difficult climate and terrain conditions. Societal adaptations—often through the use of new technologies or techniques, conquest, acculturation, or trial and error—can be found in all ecozones worldwide and at all levels of cultural, political, and economic development. The societal sum of such adaptive ingenuity would be staggering if an economic value could be placed on it. That ingenuity has led to today's societal progress by providing stepping-stones between the human settlements of past centuries and those of today.

New technological developments have also made it possible for societies to learn more about their climate conditions as well as about those in other parts of the globe. Many people, companies, and governments have learned how to put climate and weather information to a variety of uses, such as to reduce the adverse influences of climate variability and extremes on their climate activities. While learning progressed, human pressures also increased on land, air, and water resources. Population numbers increased. This in itself is not the problem; the problem with population is its *ratio* to available resources. Hence, a population density of four people per square kilometer in the West African Sahel could be seen as a problem, whereas twelve hundred people per square kilometer in a city in The Netherlands is not. The Netherlands has developed economically and politically to the point where it has overcome the constraints imposed by its climate and environment, while governments in the Sahel are still trying to do so.

Although climate's influences in various regions are being better catalogued as a result of satellite telemetry and on-the-ground research as well as information sharing via the Internet, many people still remain unaware of the obvious, let alone not so obvious, ways that climate processes and events affect socioeconomic, political, environmental, ethical, and equity issues.

A focus on the climate dimensions of social science would bring climate considerations explicitly into social science research. According to Svedin (2001), "this is not to say that the need of the natural science part of knowledge generation has diminished. But it is to say that the overall composition of the knowledge needed to address the planetary environmental issues has had to develop and grow" (p. 10).

There are several ways to bring climate dimensions to the social sciences and humanities. Some are tactical, others are strategic. Tactically, one can consider climate when it directly, visibly, and in a major way affects issues of concern to society. A strategic focus on climate runs a risk of over-focusing on the long-term issue of deep climate change at the expense of shorter term variability. However, a climate affairs approach to the climate dimension encompasses both the tactical and strategic time and space scales of concern to today's decision makers. Bringing the climate dimension to the social sciences and the humanities, and bringing the human dimensions to the natural sciences, is a win-win situation for funding agencies, researchers, and societies alike.

Problem Climates

Over forty years ago, geographer Trewartha (1961) wrote a book entitled *The Earth's Problem Climates*. The notion of "problem climates" captured my attention in 1972 as a newcomer to the research world of climate-society interactions. Trewartha's was among the first books I used when trying to understand coastal upwelling off the coast of Peru. I have since revisited and begun to reconsider the notion of "problem climates." By choice, Trewartha did not define problem climates; he was writing for science researchers and teachers and not for the general public. He noted that his book

> is designed to meet the needs of those interested in the professional aspects of climate rather than of laymen. A methodical description of all the earth's climates is not attempted, for many areas are climatically so normal or usual that they require little comment in a book which professes to emphasize the exceptional (p. 6).

From the climate perspective of 2003, is such a statement still valid? Are there any "normal climates"? Are there "problem climates"? Or, from the perspective of climate-society interactions, are there "problem societies"?

Climate is a statistical construct. It varies, fluctuates, changes, and has extremes. Before humans, climate was something with which flora and fauna had to cope. Once human settlements entered the scene, people sought to live under the climate conditions where there was some chance of survival. Aside from their human-built protection against people from other hostile settlements, they also modified their local environments to stimulate food production and maintain livestock. They did so by developing irrigation schemes, by developing ways to carry over food stuffs from one season when they were in surplus to another when they were in deficit, by developing trading practices for access to foods not of their own making, and so on. Thus, they moved into a region with what seemed, in the short term, to be good places to live.

Over time, the climate's impacts changed because of natural variability, fluctuations, and longer-term changes. As a result, people in settlements had to adjust to those changes or move on to more hospitable locations. In many instances, societies first developed water delivery systems, and in recent times, refrigeration, transportation, air conditioning or heating, and even political alliances to cope with the adverse changes in regional climate.

Problem Societies

In various locations worldwide, the physical characteristics of climate have been changing as a result of land-use decisions, for example, cutting down rainforests, increasing soil erosion, decreasing soil fertility, destroying mangroves, polluting coastal waters, land clearing for cultivation, emitting chemicals to the atmosphere, desiccating inland seas, keeping excessively large herds of animals, etc.

Policy makers need to make decisions about land use in areas with known climate-related hazards. Sometimes their decisions set up society for the impacts of varying and extreme climate and weather. Societal alterations of the environment are often found to be the underlying causes of many climate-related problems, because climate variability and its extremes serve as catalysts to climate-related disasters. Ecologist Barry Commoner (1971) made the following observation:

> The environmental crisis is a sign that the ecosphere is now so heavily strained that its continued stability is threatened. It is a warning that we must discover the source of this suicidal drive and master it before it destroys the environment—and ourselves. Environmental deterioration is caused by human action and exerts painful effects on the human condition. The environmental crisis is therefore not only an ecological problem, but also a social one (p. 109).

This, however, was not the first such assertion. Many people think of Malthus when they hear talk about the scissors-like population crisis confronting many societies—increasing numbers of people living on a dwindling natural resource base. Yet the view that increasing populations would outstrip food resources was also stated seventeen centuries ago, when Tertullian, a Carthaginian and early Christian writer around 200 A.D., made the following observation on the state of contemporary civilization:

> We find, however, in the records of the Antiquities of Man, that the human race has progressed with a gradual growth of population, either occupying different portions of the earth as aborigines, or as nomad tribes, or as exiles, or as conquerors—as the Scythians in Parthia, the Temenidae in Peloponnesus, the Athenians in Asia, the Phrygians in Italy, and the Phoenicians in Africa; or by the more ordinary methods of emigration, which they call *apaikiai* (in Greek) or colonies, for the purpose of throwing off redundant population, disgorging into other abodes their overcrowded masses. The aborigines remain still in their old settlements, and have also enriched other districts with loans of even larger populations. Surely it is obvious enough, if one looks at the whole world, that it is becoming daily better cultivated and more fully peopled than anciently. All places are now accessible, all are well known, all open to commerce; most pleasant farms have obliterated all traces of what were once dreary and dangerous wastes; cultivated fields have subdued forests; flocks and herds have expelled wild beasts; sandy deserts are sown; rocks are planted; marshes are drained; and where once were hardly solitary cottages, there are now large cities. No longer are (savage) islands dreaded, nor their rocky shores feared; everywhere are houses, and inhabitants, and settled government, and civilized life. What most frequently meets our view (and occasional complaint), is our teeming population: our numbers are burdensome to the world, which can hardly supply us from its natural elements; our wants grow more and more keen, and our complaints more bitter in all mouths,

whilst Nature fails in affording us her usual sustenance. In very deed, pestilence, and famine, and wars, and earthquakes have to be regarded as a remedy for nations, as the means of pruning the luxuriance of the human race.

(Brown, 1954; see also <www.catholicfirst.com/TheFaith/ ChurchFathers/ Volume03/Tertullian12.htm>)

Political Versus Climate Boundaries: Problem Borders

Ever since the beginning of human settlements, there have been conflicts over land-based political jurisdictions. Today's national boundaries were politically drawn as a result of conflict or accepted in order to avoid conflict. Some boundaries were drawn with natural landforms in mind—tracing rivers, forested areas, or mountains.

Commenting on the origin of borders in the United States, a geography report (NGS, 1994) noted that:

> State boundaries in the West were often created before settlement. They were drawn by people who were far away and who lacked specific information about the geography of the area. These state boundaries were imposed on the land, often following lines of latitude and longitude. Most boundaries in the East were drawn after settlement, by people who knew the land from long personal experience. Therefore, these boundaries often reflect the grain of the land—rivers, ridges, lakes (p. 11).

Atmospheric processes and climate do not respect political jurisdictions. Some settlements have favorable climate conditions on their side of a border, and others have unfavorable climate conditions. People on either side of the border have to cope with their climate-related hazards and use their climate as a resource as best they can. Not all societies, however, have the financial resources or knowledge to fully protect their citizens from climate-related hazards.

Since the advent of satellite imagery in the 1960s, we have been able to witness from space the extent to which decisions about human activities on different sides of political borders

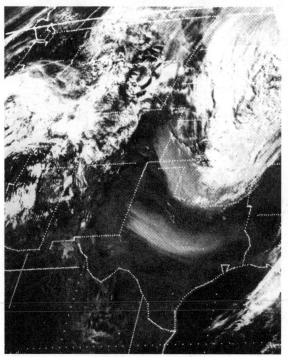

Figure 7.1. Photograph from GOES-1 geosynchronous satellite 2200 GMT (1600 CST) on 23 February 1977, showing thick dust raised by strong west and northwest winds over eastern Colorado, western Kansas and Oklahoma, and west Texas. The dust boundary is nearly coincident with the Texas–New Mexico boundary, a straight line from Hobbs northward to the Clovis-Portales area, where it laps into New Mexico about ten miles. Bright areas are clouds, except for a few areas of snow-covered mountains.

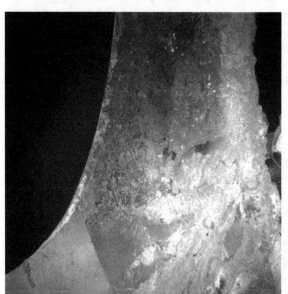

Figure 7.2. A plainly visible division between grazed and ungrazed lands in the Negev area of the Sinai desert. The photo, according to Otterman, shows the denuded, high albedo (reflectivity) regions of the Sinai and Gaza Strip in sharp contrast to the darker western Negev. The division coincides with the 1948–1949 Armistice Line between Israel and Egypt, where a fence was built in the early 1970s. (Source: NASA photo, Otterman, 1974–1977).

have different consequences for the environment (figures 7.1, 7.2). These situations illustrate how differences in land use have led to differences in land surface changes and atmospheric conditions that closely follow political jurisdictions, even under the same climate regimes.

People who live in areas at risk to the negative impacts of climate anomalies and hazards are most often the poor or politically disenfranchised. Some existing risks to natural hazards have increased because of government policies; people are allowed, if not encouraged, to live on floodplains, in low-lying coastal areas, in arid areas, and in tropical or mangrove-forested regions. Present as well as future populations are at greater risk to impacts from existing climate conditions and are likely to be more so with the advent of deep climate change.

As noted earlier, climate can be viewed as a *resource* to be exploited, a *hazard* to be avoided, and even as a *constraint* to be overcome. Yet another, often overlooked, aspect of climate is that of climate as a *scapegoat*, meaning that climate anomalies often provide decision makers with handy excuses for socioeconomic or political problems, regardless of whether those anomalies really contributed to those adverse impacts.

Thus, it is vitally important to use *all* available methods to distinguish those aspects of the impacts of a climate anomaly that can be linked legitimately to the physical aspects of climate from those that can be linked to society. Only then can policy makers take correct and appropriate action to prepare for and adapt to the adverse impacts of climate on society, and of society on climate. A failure to correctly identify the linkages between climate processes and human activities leads to policy responses that do not address the climate-related problems at hand.

The earth's problem is not just a problem with a variable and changing global climate system. It is to a large extent a problem with the way societies have chosen to develop their economies and to support their increasing populations. It is time to start pointing the finger of responsibility at problem societies rather than only at problem climates. This assertion is as valid for understanding the adverse impacts of climate variability and extremes as it is for climate change. This brings to mind a saying that appeared in a Pogo cartoon in 1971: *"I have met the enemy and he is us!"*

APPENDIX: THE TWENTIETH CENTURY'S CLIMATE EXTREMES

As is customary with the close of a century, many reviews appear about its notable events. Several retrospectives have already been produced in book and video form about the political, social, economic, and technological changes of the twentieth century. Assessments have also been undertaken by different organizations to identify the century's major disasters. Although the various final selections do not match event for event, they do overlap. The measures of relative importance vary because some chroniclers weigh loss of life as the most important consideration for ranking disasters, while others consider the spatial extent of the disaster or uniqueness of the physical damage. Individuals have a tendency to weigh events within their memory more heavily than, say, those in the earliest decades of the twentieth century. In general, though, it appears that the cost of impacts is used as a key measure for ranking importance.

NOAA's Ranking of Twentieth-Century Climate-Related Disasters

Some events, such as the 1930s Dust Bowl in the North American Great Plains, have inspired the writings of novelists and filmmakers as well as the research of physical and social scientists.

Some books about the social aspects of climate-related events include John Steinbeck's classic novel about the Dust Bowl days, *The Grapes of Wrath* (1939); Steinbeck's novel about the collapse of the California sardine fishery, *Cannery Row*; and Barry's (1998) look at the disastrous Mississippi flooding in 1927. Larson (1999) has written about the forecasting of a tropical storm in the Gulf of Mexico in the summer of 1900 that became a hurricane and made landfall at Galveston, Texas. More than 6,000 people died, and much of coastal Galveston was destroyed by the hurricane's direct hit. The following passage is from a magazine published two weeks after the 8 September Galveston hurricane and captures the spirit of the moment: "Unlike most stories of calamity, the record of this one has increased from day to day as the proof of the damage wrought has become more evident; for once the sensational papers were actually unable to exaggerate in their reports of the calamity" (*The Outlook*, 1900, p. 188).

In 1999, the U.S. National Oceanic and Atmospheric Administration (NOAA) issued its selections of what it believed were the "top weather, water and climate events of the 20th century" for the United States and for the globe. According to then-NOAA Administrator James Baker, "the lists demonstrate the wide range of weather calamities that impact much of the world's public, and supports the notion that our never-ending search to completely understand these powerful weather and climate events will remain elusive."

Aside from serving as an interesting reminder of the many ways that societies can be affected adversely by climate and weather extremes, the lists remind the reader that governments and people seem to dwell on the negative side of climate impacts. Thus, there is no comparable list by any organization that sets out to identify the most positive impacts of climate and weather anomalies of the past century.

To guide their selection, NOAA experts were asked to consider the following criteria: "an event's magnitude, meteorological uniqueness, as well as its economic impacts and death toll." Each event identified by the NOAA experts has its own history of causes, impacts, and lessons learned.

An important point to remember while going through the lists below is that these meteorological disasters generated stories not just about the death and destruction numbers but also about hope and despair, charity and compassion, sympathy and empathy for the victims, nationally as well as globally. To learn about these aspects of the worst climate-related disasters of the twentieth century, it is necessary to go to the writings of the times to capture the true spirit of the times.

NOAA's Selection of Twentieth-Century Weather- and Climate-Related Disasters in the United States (www.noaanews.noaa.gov/stories/s334.htm)

Galveston Hurricane (1900)

Hurricane Andrew (1992)

Dust Bowl (1930s)

New England Hurricane (1938)

Super Tornado Outbreak (1974)

Superstorm (1993)

Hurricane Camille (1969)

Tri-State Tornado (1925)

Great Midwest Flood (1993)

Oklahoma/Kansas Tornado Outbreak (1999)

El Niño Episodes (1982–83, 1997–98)

Great Okeechobee Hurricane and Flood (1928)

Storm of the Century (1950)

New England Blizzard (1978)

Florida Keys Hurricane (1935)

NOAA experts also collectively selected what they considered to be the major global weather, water, and climate events of the 1900s. They are not listed in any particular order.

FLOODS

- Floods usually occur in the middle and lower reaches of the major rivers in China. During this century, major flooding disasters occurred in 1900, 1911, 1915, 1931, 1935, 1950,

1954, 1959, 1991, and 1998, mainly in the Yangtze River valley.

- *Yangtze River Flood, 1931*. The summer flood along the Yangtze during July–August 1931 was the most severe, with over 51 million people affected (one-quarter of China's population). 3.7 million people perished from the greatest disaster of the century due to disease, starvation, or drowning. A prolonged drought from 1928 to 1930 preceded this flood.

- *Flood in Vietnam, 1971*. Heavy rains caused severe flooding in North Vietnam, killing 100,000.
- *Great Iran Flood, 1954*. A storm in Iran produced flooding rains, resulting in approximately 10,000 casualties.

Many of the devastating floods that occur in parts of Southeast Asia are also associated with typhoons or tropical systems.

TYPHOONS, CYCLONES, HURRICANES

- *Bangladesh Cyclone, November 1970*. The greatest tropical storm system disaster this century occurred in Bangladesh in November 1970. Winds coupled with a storm surge killed between 300,000 and 500,000 people in low-lying areas.
- *Bangladesh Cyclone 02B, April 1991*. Another cyclone struck the Chittagong region in Bangladesh in 1991, killing over 138,000 people and causing damage in excess of $1.5 billion. The tropical cyclone devastated the coastal area southeast of Dacca with winds in excess of 130 knots and a twenty-foot storm surge.
- *China Typhoons*. Several typhoons struck the eastern China coast during the early half of the century, causing tens of thousands of deaths in some storms. For example, typhoons striking China in August 1912 and August 1922 resulted in fatality counts of 50,000 and 60,000, respectively.
- *Hurricane Mitch, November 1998*. One of the strongest late-season hurricanes on record formed in the western Caribbean in October 1998. Its slow passage westward over the mountainous regions of Central America unleashed precipitation amounts as high as seventy-five inches. The resulting floods devastated the entire infrastructure of

Honduras and severely affected other countries in the region. The final estimated death toll was 11,000, the greatest loss of life from a tropical storm system in the western hemisphere since 1780.

- *Typhoon Vera, September 1958.* This typhoon's passage in 1958 caused Japan's greatest storm disaster: 5,000 deaths, 1.5 million homeless, and tremendous damage to the economy, roads, bridges, and communication systems from wind, floods, and landslides.
- *Typhoon Thelma, October 1991.* Thelma was one of the most devastating tropical systems to affect the Philippines this century. Reports indicated that 6,000 people died from catastrophic events, including dam failure, landslides, and extensive flash flooding.

DROUGHTS AND FAMINES

Numerous droughts have occurred around the globe during the past century. The impacts of droughts and famines are difficult to quantify. Their effects are devastating, and the impacts of a meteorological drought can span a couple of months or many years. Yet, as noted earlier, in most cases the occurrence of a shortfall in precipitation provides only a partial explanation for its impacts on the environment and society (e.g., Sen, 1981; Glantz, 1976). The underlying causes of famines are often related to socioeconomic factors. The most devastating droughts of the 1900s, again according to NOAA's climate experts, include the following:

- *Indian Drought of 1900.* 250,000–3.25 million perished as a result of drought, starvation, and disease.
- *Chinese Famine of 1907.* More than 24 million died from starvation.
- *Chinese Famine of 1928–1930.* More than 3 million perished in northwest China.
- *Chinese Famine of 1936.* Five million Chinese died in what was called the New Famine.
- *Chinese Drought of 1941–42.* More than 3 million perished from starvation.
- *Indian Drought of 1965–67.* More than 1.5 million died in India.

- *Soviet Union (Ukraine and Volga regions), 1921–22.* 250,000–5 million perished.
- *Sahel Drought.* Famines and droughts have occurred in the West African Sahel in 1910–14, 1940–44, 1970–85. In 1972–75 and again in 1984–85, which claimed over 600,000 lives.

TORNADOES

- The United States is considered the tornado capital of the world, with more tornadoes annually than any other country. Two notable outbreaks included the "Super Tornado Outbreak of 1974" (315 deaths) and the "Tri-State Tornado of 1925" (695 deaths).

WINTER STORMS/ BLIZZARDS

- *Iran Blizzard, February 1972.* A blizzard in Iran in February 1972 ended a four-year drought, but the weeklong cold snap and snow left approximately 4,000 people dead.
- *Europe Storm Surge, 1953.* One of Europe's greatest natural disasters occurred during January–February 1953. Violent winter storms caused storm surges, which caused flooding in areas of the Netherlands and the United Kingdom. Almost 2,000 people perished.

DEADLY SMOG

- *London's Great Smog, December 1952.* Stagnant air caused by an inversion combined with industrial and residential emissions to create an air pollution episode without parallel in this century. Four thousand deaths were attributed to the poisonous air; four thousand additional fatalities occurred due to related causes.

THE 1982–83 EL NIÑO

The economic impacts of the 1982–83 El Niño were huge. Along the west coast of South America, the fishing industries in Ecuador and Peru suffered heavily when their anchovy harvest failed and sardines unexpectedly moved south into Chilean waters. However, wild shrimp populations expanded sharply off the coasts of Peru and Ecuador. Also, changed atmospheric circulation patterns steered tropical systems off their usual tracks to islands such as Hawaii and Tahiti, which are unaccustomed to such severe weather. They also caused the monsoon rains to fall over the central Pacific instead of the western Pacific. The lack of rain in the western Pacific led to droughts and disastrous forest fires in Indonesia and Australia. Winter storms battered southern California and caused widespread flooding across the southern United States, while unusually mild weather and a lack of snow was evident across much of the central and northeastern United States. Overall, the loss to the world economy in 1982–83 as a result of the changes in climate due to El Niño amounted to over $8 billion. The toll in terms of human suffering is much more difficult to estimate (OGP, 1994). The *U.S. News & World Report* estimated $21 billion 1997 U.S. dollars (about $13 billion in 1982–83 dollars) (Brownlee and Tangley, 1997).

A NOAA survey (Sponberg, 1999) of global impacts caused by the 1997–98 El Niño estimated global damages at $25–$33 billion, whereas Munich Re, a reinsurance corporation, estimated this El Niño's impacts at $96 billion (Munich Re, 2000). As one can see, estimating El Niño or other climate-related damages is not a precise science and involves a lot of "guesstimates."

The WMO and Other Reviews
of the Twentieth Century's Climate

In 1995, the United Nation's World Meteorological Organization (WMO) was authorized by its member states to undertake a comprehensive, detailed, popular review of the climate history of the twentieth century. The review was to involve physical science as well as societal impacts of short-term weather, climate variability, and climate change. Global in scope, the review is a comprehensive and authoritative collection of facts and stories about how weather and climate averages, anomalies, and changes have affected human activities. It also discusses how human activities have influenced climate behavior. The review (Burroughs, 2003) addresses ways scientists have improved the monitoring, modeling, and prediction of climate behavior, and how societies have or could use that information to cope with future changes.

Canada's environment ministry identified the top weather events in Canada in the twentieth century (Environment Canada, 2001). Australia's Bureau of Meteorology, too, has identified climate extremes of the twentieth century (BOM, 2000). Several other countries and some individual states within the United States have also sought to identify their top weather and climate events.

It would be useful for comparative purposes to convince each government in the United Nations to ask its scientific experts to identify the country's remarkable climate- and weather-related events of the twentieth century.

REFERENCES

AccuWeather, 2001. Tornado frequency in the United States, 1950–2000 (www.accuweather.com).

Alexander, T., 1974. Ominous change in the weather. *Fortune Magazine* (February), 90–152.

Ali, M. M., ed. 1998. *Bangladesh floods: Views from home and abroad.* Dhaka: University Press.

Allen, B. J., 1993. The problems of upland land management. In *South-East Asia's environmental future: The search for sustainability,* ed. H. Brookfield and Y. Byron, 225–37. Tokyo: United Nations University Press.

Anderson, D., and R. Grove, eds. 1987. *Conservation in Africa: People, policies and practice.* Cambridge, UK: Cambridge University Press.

Athanasiou, T., and P. Baer, 2001. Climate change after Marrakech: Should environmentalists still support the Kyoto Protocol? *FPIF (Foreign Policy in Focus) Discussion Paper #5,* December (www.fpif.org/papers/marrakesh_body.html).

Bailey, R., 1999. Precautionary tale: The latest environmentalist concept—the precautionary principle—seeks to stop innovation before it happens. *Reason Magazine,* April (reason.com/9904/fe.rb.precautionary.shtml).

Ball, J., 2002. US joins fight against California clean-air effort. *Wall Street Journal,* 10 October.

Bandyopadhyaya, J., 1983. *Climate and world order: An inquiry into the natural causes of underdevelopment.* New Delhi, India: South Asian Press.

Barnston, A. G., M. H. Glantz, and Y. He, 1999. Predictive skill of statistical and dynamical climate models in SST forecasts during the 1997–98 El Niño episode and the 1998 La Niña onset. *Bulletin of the American Meteorological Society* 80(2):217–42.

Barr, J., 2001. Papua New Guinea. In *Once burned, twice shy? Lessons learned from the 1997–98 El Niño,* ed. M. H. Glantz, 159–75. Tokyo: United Nations University Press.

Barris, L., 1999. It's time to get on the CALFED bandwagon. *Chico Examiner,* 29 December (ww.chicoexaminer.com/Columns/col_LB/col_LB_19991229.html).

Barry, J. M., 1998. *Rising tide: The great Mississippi flood of 1927 and how it changed America.* New York: Simon & Schuster.

BBC News Online, 2000. Climate talks end in failure. 25 November (news.bbc.co.uk/low/English/sci/tech/neurid_1040000/1040091.htm).

Begley, S. (with M. Miller and M. Hager), 1988. The endless summer? *Newsweek,* 11 July, 18–20.

Benedick, R. E., 1991. *Ozone diplomacy: New directions in safeguarding the planet.* Cambridge: Harvard University Press.

Berk, R., 1994. Public perception of global warming. In *Anticipating global change surprises,* eds. S. H. Schneider and B. L. Turner II. Aspen, CO: Aspen Global Change Institute (www.gcrio.org/ASPEN/science/eoc94/EOC2/EOC2-2.html).

Betsill, M. M., M. H. Glantz, and K. Crandall, 1997. Preparing for El Niño: What role for forecasts? *Environment* 39(10):6–13, 26–30.

Biodiversity Clearing-House Mechanism, 2001. Biodiversity frequently asked questions. Brussels, Belgium: Convention on Biological Diversity (bch-cbd.naturalsciences.be/belgium/biodiversity-faq.htm).

Bjerknes, J., 1969. Atmospheric teleconnections from the equatorial Pacific. *Monthly Weather Review* 97:163–72.

Blackwelder, B., 2002. Bush administration: Tools for environmental destruction. *Friends of the Earth Newsmagazine,* 32, p. 2 (www.foe.org/res/pubs/news2002.html).

BOM (Bureau of Meteorology), 2000. *Australian climate extremes of the 20th century.* Australia: BOM (www.bom.gov.au/climate/c20thc/index.shtml).

Brookes, W., 1990. White House effect vs. greenhouse effect. *The Washington Times,* p. f-2.

Brooks, C. E. P., 1926. *Climate through the ages: A study of the climatic factors and their variations.* Second Revised Edition in 1970. New York: Dover Publications.

Brown, B. G., 1988. Climate variability and the Colorado River Compact: Implications for responding to climate change. In *Societal responses to regional climate change: Forecasting by analogy,* ed. M. H. Glantz, 293. Boulder, CO: Westview Press.

Brown, L. R., and E. P. Eckholm, 1974. *By bread alone.* New York: Praeger.

Brown, H., 1954. *The challenge of man's future.* New York: The Viking Press, Inc.

Brownlee, S., and L. Tangley, 1997. The wrath of El Niño. *US News & World Report,* 6 October (rossby.metr.ou.edu/~spark/AMON/v1_n5/News/elnino/USNEWS_wrath.html).

Brunner, B., ed. 2001. *TIME almanac 2002.* Boston: Family Education Company, 625.

Bryson, R. A., and T. J. Murray, 1977. *Climates of hunger: Mankind and the world's changing weather.* Madison: University of Wisconsin Press.

Budyko, M. I., 1988. Anthropogenic climate changes. Paper presented at the World Congress on Climate and Development, 7–10 November, Hamburg, Germany.

Burke, D., A. Carmichael, D. Focks et al., 2001: *Under the weather: Climate, ecosystems, and infectious disease.* Committee on Climate, Ecosystems, Infectious Diseases, and Human Health: Board on Atmospheric Sciences and Climate, 113–14. Washington, D.C.: National Research Council.

Burroughs, W., ed. 2003. *Climate: Into the 21st century.* Cambridge, U.K.: Cambridge University Press.

Burroughs, W. J., 1997. *Does the weather really matter? The social implications of climate change.* Cambridge, U.K.: Cambridge University Press.

CBD (Convention on Biological Diversity), 2002. Climate change and biodiversity: Overview of the interlinkages between biological diversity and climate change—The climate change phenomenon (www.biodiv.org/programmes/crosscutting/climate/interlinkages.asp).

CCX (Chicago Climate Exchange), 2002. NASD and the Chicago Climate Exchange reach historic agreement. Chicago: CCX (www.chicagoclimatex.com).

CDC (Climate Diagnostics Center), 2002. Multivariate ENSO Index time series. Boulder: University of Colorado (www.cdc.noaa.gov/~kew/MEI).

CDREE (Canadian Department of Regional Economic Expansion),

1978. *Regional economic expansion.* Discussion paper on drought in western Canada. Ottawa, Ontario: CDREE.

CENR (Committee on Environment and Natural Resources), 2001. *Ecological forecasting: Agenda for the future.* Reston, Va: NBII National Program Office (www.nbii.gov).

Chambers, R., R. Longhurst, D. Bradley, and R. Feachem, 1979. *Seasonal dimensions to rural poverty: Analysis and practical implications.* Brighton, U.K.: Institute of Development Studies, University of Sussex.

Changnon, S. A., ed., 1996. *The great flood of 1993: Causes, impacts, and responses.* Boulder, Colo.: Westview Press.

Changnon, S. A., 1994. *The Lake Michigan diversion at Chicago and urban drought: Past, present and future regional impacts and responses to global climate change.* Boulder, Colo.: University Corporation for Atmospheric Research.

Charney, J. G., 1975. Dynamics of deserts and drought in the Sahel. *Quarterly Journal of the Royal Meteorological Society* 101:193–202.

Chasek, P. S., 2001. *Earth negotiations: Analyzing thirty years of environmental diplomacy.* Tokyo, Japan: United Nations University Press, p. 4.

Chu Co-Ching, 1954. Climatic change during historic times in China. In: *Collected scientific papers: Meteorology 1919–1949,* 272. Peking: Academia Sinica.

Climatic Change, 1996. Special themes issue: Integrated assessment. *Climatic Change* 34(3–4):315–566 (www.kluweronline.com/issn/01650009).

Columbia Encyclopedia, 2000. Iran hostage crisis. Sixth edition. New York: Columbia University Press.

Colwell, R. R., and J. A. Patz, 1998. *Climate, infectious disease and health: An interdisciplinary perspective.* Washington, D.C.: American Academy of Microbiology.

Commoner, B., 1971. *The closing circle: Nature, man, and technology.* New York: Alfred A. Knopf.

Conservation International, 2002. Washington, D.C.: Conservation International (www.conservation.org).

Cox, J. H., 2002. *Stormchasers: The turbulent history of weather prediction from Franklin's kite to El Niño.* New York: John Wiley & Sons, Inc.

Cox, G. W., 1875. *The Crusades: Epochs of history.* New York: Scribner, Armstrong, and Co.

Crossette, B., 1999. US study funds lack of control in UN food aid to

North Korea. *New York Times,* 12 October (www.globalpolicy.org/socecon/un/food.htm).

Crowley, T. J., 1990. Are there any satisfactory geological analogs for a future greenhouse warming? *Journal of Climate,* 3:1282–92.

Crutzen, P., 1995. My life with O, NO_x and other YZO_xs. *Les Prix Nobel,* 123–57. Stockholm: Almqvist and Wiksell International.

CSTPR (Center for Science and Technology Policy Research), 2002. *Weather and climate forecast use and value bibliography.* Boulder, Colo.: CSTPR (sciencepolicy.colorado.edu/biblio).

Daley, S., 2002. Europeans give Bush plan on climate change a tepid reception. *New York Times,* 15 February (www2.uwsuper.edu/acaddept/hps/mjohnson/200/EuroReactBG.htm).

D'Arge, R. C., 1979. Climate and economic activity. In *World climate conference: A conference of experts on climate and mankind,* 302–17. Geneva: World Meteorological Organization.

Dasmann, R. F., 1959. *Environmental conservation.* New York: John Wiley & Sons.

Davis, M., 2001: *Late Victorian holocausts: El Niño, Famines, and the making of the Third World.* New York: Verso Books.

DePalma, A., 2002. Cuban cash reopens food trade. Washington, D.C.: Center for International Policy (www.ciponline.org/cuba/news articles/nyt021402depalma.htm).

Deudney, D., 1990. The case against linking environmental degradation and national security. *Millennium* 19:461–76.

Diamond, J., 1997. *Guns, germs, and steel: The fates of human societies,* New York: W.W. Norton & Co., Inc.

Diaz, H. F., and F. Markgraf, eds. 1992: *El Niño: Historical and paleoclimatic aspects of the Southern Ocean.* Cambridge, U.K.: Cambridge University Press.

Drillbits & Tailings, 2000. Russian gas companies follow receding arctic ice. *Drillbits & Tailings,* 5(15):19 September. Berkeley, CA: Project Underground (www.moles.org/ProjectUnderground/drillbits/5_15/1.html).

Earthbeat, 2002. Scientists say you can't drought proof Australia. Broadcast of Saturday, 12 December. New South Wales, Australia: Radio National (www.abc.net.au/rn/science/earth/stories/s695724.htm).

Eco, 2002. The Climate Action Network conference newsletter from the international negotiations on climate change under the UNFCCC.

Washington, D.C.: Climate Action Network (www.climate network.org/eco/).

Ehrlich, P. R., C. Sagan, D. Kennedy, and W. O. Roberts: 1984: *The cold and the dark: The world after nuclear war.* New York: W.W. Norton & Co., Inc.

El Universo, 1997. Según congresista peruano: El Niño impedirá conflicto. 26 October, p. 3.

Emanuel, K. A., K. Speer, R. Rotunno et al., 1995. Hypercanes: A possible link in global extinction scenarios. *Journal of Geophysical Research-Atmospheres* 100(D7):13,755–65.

Environment Canada, 2002. *Environment Canada's green lane: Educational resource guides.* Primer on Climate Change (www.ec.gc.ca/edu_e.html).

Environment Canada, 2001. *Top weather events of the 20th century.* Hull, Quebec: Environment Canada (www.ec.gc.ca/press/vot20_f_e.htm).

Environment Canada, 1998. *Ice storm '98.* North York, Ontario: Environment Canada.

ESIG (Environmental and Societal Impacts Group), 2001. Newspaper headlines from various countries attacking the Bush administration policy to abandon the Kyoto process. Boulder, Colo.: National Center for Atmospheric Research.

Fagan, Brian M., 2000: *Floods, famines and emperors: El Niño and the fate of civilizations.* New York: HarperCollins.

Farvar, M. T., and J. R. Milton, eds. 1972. *The careless technology: Ecology and international development.* Garden City, N.Y.: Natural History Press.

Feenstra, J. F., I. Burton, J. B. Smith, and R. S .J. Tol, 1998. *Handbook on methods for climate change impact assessment and adaptation strategies.* Version 2.0. Amsterdam: Institute for Environment Studies.

Fialka, J. J., 2002. White House backs Indian challenger to head panel probing climate change. Seattle, WA: *Pressure Point* (www.pressure point.org/pp_news_exxon_wsj_4_4_02.html).

Fischer, D. H., 1996. *The great wave: Price revolutions and the rhythm of history.* New York: Oxford University Press.

Fischhoff, B., 1994. What forecasts (seem to) mean. *International Journal of Forecasting,* 10, 387–403.

Flannery, T. F., 1995. *The future eaters: An ecological history of the Australasian island and people.* New York: George Braziller, Inc.

Forest Monitor, 2001. *Sold down the river: The need to control transnational forestry corporations: A European case study.* Cambridge, U.K.: Forest Monitor Ltd., March.

Freeman, J. W., 2001. *Storms in space.* Cambridge, U.K.: Cambridge University Press.

GCI (Global Commons Institute), 1996. GCI: Defending the value of human life. London: Global Commons Institute (www.gci.org.uk/vol/vol.html).

GECP (Global Environmental Change Programmes), 2001. *Global change and the earth system: A planet under pressure.* IGBP Science No. 4. Stockholm: International Geosphere-Biosphere Programme Secretariat.

Ghil, M. and S. Childress, 1987. *Topics in geophysical fluid dynamics: Atmospheric dynamics.* Berlin: Springer-Verlag.

Gifis, S., 1991. *Law dictionary.* Third Edition. New York: Barron's Educational Series, Inc., 195–96.

Gill, A. E., and E. M. Rasmusson, 1983. The 1982–83 climate anomaly in the equatorial Pacific. *Nature* 306:229–34.

Glantz, M. H., 2001a. *Currents of change: El Niño and La Niña impacts on climate and society.* Cambridge, U.K.: Cambridge University Press.

Glantz, M. H., ed. 2001b. *Once burned, twice shy? Lessons learned from the 1997–98 El Niño.* Tokyo: United Nations University Press.

Glantz, M. H., 2000. Climate-related disaster diplomacy: A US-Cuban case study. *Cambridge Review of International Affairs* 14(1):233–53.

Glantz, M. H., ed. 1999. *Creeping environmental problems and sustainable development in the Aral Sea Basin.* Cambridge, U.K.: Cambridge University Press.

Glantz, M. H., ed. 1994a. *Drought follows the plow: Cultivating marginal areas.* Cambridge, U.K.: Cambridge University Press.

Glantz, M. H., 1994b. *The impacts of climate on fisheries.* UNEP Environment Library No. 13. Nairobi: United Nations Environment Programme.

Glantz, M. H., ed. 1992. *Climate variability, climate change, and fisheries.* Cambridge, U.K.: Cambridge University Press.

Glantz, M. H., 1990. *On assessing winners and losers in the context of global warming.* Report of workshop held 18–21 June 1990 in St. Julians, Malta. Boulder, Colo.: Environmental and Societal Impacts Group/National Center for Atmospheric Research.

Glantz, M. H., ed. 1988: *Societal responses to regional climatic change: Forecasting by analogy.* Boulder, Colo.: Westview Press.

Glantz, M. H., and R. W. Katz, 1985. Drought as a constraint to development in Sub-Saharan Africa. *Ambio* 14(6):334–39.

Glantz, M. H., 1982. Consequences and responsibilities in drought forecasting: The case of Yakima 1977. *Water Resources Research* 18(1):3–13.

Glantz, M. H., 1979. The science, politics and economics of the Peruvian anchoveta fishery. *Marine Policy* 3(3):201–10.

Glantz, M. H., ed. 1977. *Desertification: Environmental degradation in and around arid lands.* Boulder, Colo.: Westview Press.

Glantz, M. H., 1976. Nine fallacies of natural disaster: The case of the Sahel. In *Politics of natural disaster: The case of the Sahel drought,* ed. M. H. Glantz, 3–24. New York: Praeger.

GLOBE (Global Legislators Organization for a Balanced Environment) Southern Africa, 2000. *Climate change in Africa.* Cape Town, South Africa (www.globesa.org).

Glossop, R., 1999. The last century and the next. Published on the Internet by the First Unitarian Church of Alton, Ill. (www.geocities.com/Athens/Forum/1718/ss120599.html).

Glover, D., and T. Jessup, 1999. *Indonesia's fires and haze: The cost of catastrophe.* Singapore: Institute of Southeast Asian Studies.

Glynn, P. W., 1990. *Global ecological consequences of the 1982–83 El Niño-Southern Oscillation.* Amsterdam: Elsevier.

Gray, W. M., J. D. Sheaffer, and C. W. Landsea, 1997. Climate trends associated with multidecadal variability of Atlantic hurricane activity. In *Hurricanes,* ed. H. Diaz and R. S. Pulwarty. New York: Springer.

Greenpeace, 1997. *El Niño and climate change: Troubled waters.* Amsterdam: Greenpeace International Report.

Greenwood, D., 1957. Preface. In *Climate and economic development in the tropics* ed. D. H. K. Lee. New York: Harper and Brothers.

Gribbin, J., 1976. *Forecasts, famines and freezes: Climate and man's future.* London: Wildwood House Publishers.

Grobecker, A. J., S. C. Coroniti, and R. H. Cannon, Jr., 1974. *The report of findings: The effects of stratospheric pollution by aircraft,* COT-TST-75-50. Springfield, VA: U.S. Department of Transportation, Climatic Impact Assessment Program, National Technical Information Service.

Gupta, J., 2001. *Our simmering planet: What to do about global warming?* London: Zed Books Ltd.

Haimson, L., 2002. This just in. Seattle, WA: *Grist Magazine* (www.gristmagazine.com/heatbeat/thisjustin112202.asp).

Halperin, D., 2002. Black Christmas: The Australian bushfires of 2001. *American Forests,* 108(2):18–22.

Halpern, S., 2001. *Four wings and a prayer: Caught in the mystery of the Monarch butterfly.* New York: Pantheon Books.

Hansen, J. E., 1988. The greenhouse effect: Impacts on current global temperature and regional heat waves. *Testimony at hearing on the greenhouse effect and global climate change,* Committee on Energy and Natural Resources, U.S. Senate, One Hundredth Congress, First Session. Washington, D.C: U.S. Government Printing Office.

Hare, F. K., 1978. *The restless atmosphere.* New York: Harper & Row.

Harper, T., 2002. Premiers ambush PM on climate deal: Klein stuns Chrétien with anti-Kyoto letter, February 16, *Toronto Star.*

Harremoës, P., D. Gee, M. MacGarvin et al., 2002. *The precautionary principle in the 20th century: Late lessons from early warnings.* London: Earthscan Publications Ltd.

Hinman, R., 1888. *Eclectic physical geography.* New York: American Book Company.

Hoff, B., and C. Smith III, 2000. *Mapping epidemics: A historical atlas of disease.* Danbury, Conn.: Grolier Publishing.

Homer-Dixon, T. F., 1999. *Environment, scarcity, and violence.* Princeton, N.J.: Princeton University Press.

Homer-Dixon, T. F., 1991. On the threshold: Environmental changes as causes of acute conflict. *International Security* 16(2):76.

Hope, K., 2001. Greece and Turkey: The Balkan powers continue their complicated dance of diplomacy. *Europe,* 404, p. 12.

Houghton, J. T., Y. Ding, D. J. Griggs, M. Noguer et al., 2001. *Climate change 2001: The scientific basis.* Cambridge, U.K.: Cambridge University Press.

Huber, P., 2000. *Hard green: Saving the environment from the environmentalists: A conservative manifesto.* New York: Basic Books.

Huntington, E. [1915] 1971. *Civilization and climate.* Hamden, Conn.: The Shoestring Press.

ICIHI (Independent Commission for International Humanitarian Issues), 1985. *Famine: A man-made disaster.* Report for the ICIHI. London: Zed Books.

Impact Team, 1977. *The weather conspiracy: The coming of the new Ice Age.* New York: Ballantine Books.

IPCC (Intergovernmental Panel on Climate Change), 2001. *Climate change 2001: Impacts, adaptation, and vulnerability.* Contribution of Working Group II to the Third Assessment Report of the IPCC. Cambridge, U.K.: Cambridge University Press.

IPCC, 1996. *Climate change 1995: The science of climate change,* ed. J. T. Houghton, L. G. Meira-Filho, B. A. Callander, N. Harris, A. Kattenberg, and K. Maskell. Contribution of Working Group II to the Second Assessment Report of the IPCC. Cambridge, U.K.: Cambridge University Press.

IPCC, 1990. *Climate change: The IPCC scientific assessment,* ed. J.T. Houghton, G. J. Jenkins, and J. Ephraums. Cambridge, U.K.: Cambridge University Press.

Jamieson, D., 1988. Grappling for a glimpse of the future. In *Societal responses to regional climatic change: Forecasting by analogy,* ed. M. H. Glantz, 73–94. Boulder, Colo.: Westview Press.

Japan Times, 2001. Utilities eye "weather derivative" deal. *Japan Times Online,* November 22 (www.japantimes.co.jp/cgibin/getarticle.pl5?nb20011122b2.htm).

Jomini, L.G., 1820. *Histoire critique et militaire des guerres de la révolution. Tome Sixième, Campagne de 1794–second Période.* Book 7, Vol. 6, Chapter 152, p. 208. Paris: Anselin et Pochard.

Kamarck, A. M., 1976. *The tropics and economic development: A provocative inquiry into the poverty of nations.* World Bank Report. Baltimore, Md.: The Johns Hopkins University Press for World Bank.

Kates, R., J. Ausubel, and M. Berberian, 1985. *Climate impact assessment: Studies of interaction of climate and society.* SCOPE 27. New York: John Wiley & Sons.

Klimm, L. E., O. P. Starkey, and N. F. Hall, 1940. *Introductory economic geography.* Second edition. Orlando, Fla.: Harcourt Brace & Co.

Kluckhohn, C. and F. L. Strodtbeck, [1961] 1973. *Variations in value orientations.* Westport, Conn.: Greenwood Press.

Kondratyev, K. Ya., 1988. *Climate shocks: Natural and anthropogenic.* New York: John Wiley & Sons.

Ladurie, Emmanuel Le Roy, 1971. *Times of feast, times of famine: A history of climate since the year 1000.* Garden City, N.Y.: Doubleday & Company.

Lamb, Hubert H., [1982] 1995. *Climate, history and the modern world.* London: Methuen & Co. Ltd.

Landsberg, H. E., 1956. The climate of towns. In *Man's role in changing the face of the earth*, ed. W. L. Thomas, Jr., 584. Chicago: University of Chicago Press

Larson, Eric, 1999. *Isaac's storm*. New York: Crown Publishers.

Lewis, J., 1999. *Development in disaster-prone places: Studies of vulnerability*. London: Intermediate Technology Publications.

Linden, E., 1998. *The future in plain sight: Nine clues to the coming instability*. New York: Simon and Schuster.

Lippmann, W., [1955] 1989. *The public philosophy*. New Brunswick, N.J.: Transaction Publishers.

Litfin, K. T., 1994. *Ozone discourses: Science and politics in global environmental cooperation*. New York: Columbia University Press.

Ljunggren, D., 2002. Canadian Parliament backs Kyoto ratification plan. Berkeley, CA: Environmental News Network.

Lofchie, M., 1975. Political and economic origins of African hunger. *Journal of Modern African Studies*, 13(4), 551–567.

Markham, S. F., 1947. *Climate and the energy of nations*. London: Oxford University Press.

Martin, B., 1979. *The bias of science*. O'Connor, Australia: Society for Social Responsibility.

Mattingly, G., 1959. *The Armada*. Boston: Houghton Mifflin.

McFadden, R. D., 2001. Imperfect storm is less of a blow than was feared. *New York Times*, 6 March (www.nytimes.com/learning/general/featured_articles/010307wednesday.html).

McPhee, J., 1967. *Oranges*. New York: Farrar, Straus & Giroux.

Meinke, H., W. Wright, P. Hayman, and D. Stephens, 2003. Managing cropping systems in variable climates. In Principles of field crop production, 26–77, ed. J. Pratley. Melbourne: Oxford University Press.

Miller, K. A., G. R. Munro, T. L. McDorman, R. McKelvey, and P. Tyedmers, 2001. *The 1999 Pacific Salmon Agreement: A sustainable solution?* Occasional Paper No. 47. Orono, ME: Canadian-American Center, University of Maine.

Miller, K. A., and M. H. Glantz, 1988. Climate and economic competitiveness: Florida freezes and the global citrus processing industry. *Climatic Change*, 12, 135–164.

Molina, M. J., and F. S. Rowland, 1974. Stratospheric sink for chlorofluoromethanes: Chlorine atomic-catalyzed destruction of ozone. *Nature* 274:810–12.

Monnik, K., 2001. Role of drought early warning systems in South

Africa's evolving drought policy. In *Early warning systems for drought preparedness and drought management*, ed. D. A. Wilhite, M. V. K. Sivakumar, and D. A. Wood. Proceedings of Expert Meeting held 5–7 September 2000 in Lisbon, Portugal. Geneva: WMO (www.drought.unl.edu/monitor/EWS/).

Mortimore, M., 1989. *Adapting to drought: Farmers, famines and desertification in West Africa.* Cambridge, U.K.: Cambridge University Press.

Mundle, R., 2000. *Fatal storm: The inside story of the tragic Sydney-Hobart race.* New York: McGraw-Hill.

Munich Re, 2000. *Topics 2000: Natural catastrophes—The current position.* Munich, Germany: Munich Re Geoscience Research Group.

Murphy, R.C., 1954. The guano and the anchoveta fishery. Reprinted in: *Resource management and environmental uncertainty: Lessons from coastal upwelling,* ed. M. H. Glantz and J. D. Thompson, 81–106. New York: John Wiley & Sons.

Myers, N. and J. Kent, 1995. *Environmental exodus: An emergent crisis in the global arena.* Washington, D.C.: The Climate Institute.

Najam, A., 2000. Future directions: The case for a "Law of the Atmosphere." *Atmospheric Environment,* 34(33):4047–49.

NAS (National Academy of Sciences), 1999. Space weather: A research perspective: What is space weather? Washington, D.C.: Space Studies Board, NAS (www.nas.edu/ssb/swwhat.html).

National Assessment Synthesis Team, 2000. *Climate change impacts on the United States: The potential consequences of climate variability and change.* Washington, D.C.: U.S. Global Change Research Program.

Newell, F., 1896. Irrigation on the Great Plains. *US Department of Agriculture, Yearbook of agriculture 1896.* Washington, DC: Government Printing Office, 167–196.

NGS (National Geography Standards), 1994. Washington, D.C.: National Geographic, Division of Research and Exploration.

Nicholls, N., 1986. A method for predicting Murray Valley encephalitis in southeast Australia using the Southern Oscillation. *Australian Journal of Experimental Biological and Medical Science,* 64, 587–94.

Nicholls, N., and T. S. Kestin, 1998. Communicating climate. *Climatic Change,* 40, 417–20.

NMU (Northern Michigan University), 2000. Northern Michigan University psychology definition (www.nmu.edu/psychology/defin.htm).

NOAA (National Oceanic and Atmospheric Administration), 2001. Billion-dollar weather disasters. Washington, D.C.: NOAA (www.noaa.gov).

NRC (National Research Council), 2002. *Abrupt climate change: Inevitable surprises.* Washington, D.C.: Committee on Abrupt Climate Change (www.nap.edu/catalog/10136.html).

NRL (Naval Research Laboratory), 2003. Space weather prediction: Research and development. Washington, D.C.: Plasma Physics Division, NRL (ppdweb.nrl.navy.mil/whatsnew/prediction/).

NWS (National Weather Service), 1994. *Superstorm of March 1993: March 12–14 1993.* Natural Disaster Survey Report. Washington, D.C.: U.S. Department of Commerce.

Offut, C., 2002. Dying monarchs. *E-Magazine* 13(3):25 (www.emagazine.com/may-june_2002/_0502contents.html).

Ogola, J. S., 1997. *Potential impacts of climate change in Kenya.* Nairobi: Climate Network Africa (lion.meteo.go.ke/cna/).

OGP (Office of Global Programs), NOAA, 1994. *Reports to the nation on our changing planet: El Niño and climate prediction.* Silver Spring, Md.: Office of Global Programs. Out of print (www.ogp.noaa.gov/library/index.htm).

Oliver, J., N. R. Britton and M. K. Hannes, 1984. The Ash Wednesday bushfires in Victoria: 16th February 1983. *Disaster Investigation Report No. 7.* Queensland: James Cook University.

Olori, T., 2002. Nigeria: Desertification threatens economy, food security. *ProutWorld News* (www.proutworld.org/news/en/2002/aug/20020818des.htm).

Ominde, S. H. and C. Juma, eds. 1991. *A change in the weather: African perspectives on climatic change.* Nairobi: African Centre for Technology Studies.

Overpeck, J. T., D. Rind, R. Healy, and A. Lacis, 1996. Possible role of dust-induced regional warming in abrupt climate change during the last glacial period. *Nature* 384:447–49.

Pearce, F., 2002. Africans go back to the land as plants reclaim the desert. *New Scientist* 175(2361):4–5.

PEER (Public Employees for Environmental Responsibility), 2001. Wise use. Washington, D.C.: PEER (www.peerorg/wise_use/).

Pfaff, A., K. Broad, and M. H. Glantz, 1999. Who benefits from climate forecasts? *Nature* 397:645–46.

Pianin, E., 2000. Deal on Cuba would ease U.S. sanctions: Food, medicine sales allowed. *The Washington Post* (28 June):A1.

Pielke, Jr., R. A., 1995. *Hurricane Andrew in South Florida: Mesoscale weather and societal responses.* Boulder, Colo.: National Center for Atmospheric Research.

Pielke, Jr., R. A., and D. Sarewitz, 2000. Global warming anyone? *The Washington Times,* 2 February.

Ponte, L., 1977. When the Sahel freezes over. *Skeptic,* 37–58.

Ponte, L., 1976. *The cooling.* Englewood Cliffs, N.J.: Prentice-Hall, Inc.

Potok, C., 2001. The war doctor. In *Old men at midnight.* New York: Alfred A. Knopf Publishers.

Quote Garden, 2003. Weather quotes (www.quotegarden.com).

Rachlinski, J. J, 2000. The psychology of global climate change. *University of Illinois Law Review,* 1.

Rasmusson, E. M., and T. H. Carpenter, 1982. Variations in tropical sea surface temperature and surface wind fields associated with the Southern Oscillation/El Niño. *Monthly Weather Review* 110:354–84.

RCGP (Royal College of General Practitioners), 2001. Editorial by Dr. Mike Thompson. *RCGP International Newsletter,* 27 (www.rcgp.org.uk/rcgp/international/newsletters/jun01/).

Redclift, M., 1987. *Sustainable development: Exploring the contradictions.* New York: Routledge Press.

Redfield, W. C., 1855. Cape Verde and Hatteras hurricane, and other storms, with a hurricane chart. *AAAS (American Association for the Advancement of Science) Proceedings of the Eighth Meeting, May 1854,* 208–15. New York: G. P. Putnam & Co.

Revkin, A. C., 2002. Bush offers plan for voluntary measures to limit gas emissions. *The New York Times on the Web.* 15 February (www.nytimes.com).

Riebsame, W. E., 1988. *Assessing the social implications of climate fluctuations: A guide to climate impact studies.* Nairobi: United Nations Environment Program.

Risbey, J., M. Kandlikar, and A. Patwardhan, 1996. Assessing integrated assessments. *Climatic Change* 34(3–4):369–95.

Rosenzweig, C., and D. Hillel, 1998. *Climate change and the global harvest: Potential impacts of the greenhouse effect on agriculture,* 261–62. New York: Oxford University Press.

Rosenzweig, C., and W. D. Solecki, eds. 1999. *Climate change and a global city: The potential consequences of climate variability and change, Metro East Coast.* Report for the U.S. Global Change

Research Program, National Assessment of the Potential Consequences of Climate Variability and Change for the United States. New York: Columbia Earth Institute.

Roszak, T., M. E. Gomes, and A. D. Kanner, 1995. *Ecopsychology: Restoring the earth, healing the mind.* San Francisco, Sierra Club Books.

Rotmans, J., and M. van Asselt, 2001. *Uncertainty in integrated assessment modeling: A labyrinthic path.* The Netherlands: Kluwer Academic Publishers.

Rotmans, J., and M. van Asselt, 1996. Integrated assessment: A growing child on its way to maturity. *Climatic Change* 34(3–4):327–36.

Saarinen, T. F., 1966. *Perception of the drought hazard on the Great Plains.* Research Paper No. 106. Chicago: University of Chicago.

Sachs, J., A. D. Mellinger, and J. L. Gallup, 2001. The geography of poverty and wealth. *Scientific American* 284(3):70–75.

Sahn, David E., ed. 1989. *Seasonal variability in third world agriculture: The consequences for food security.* Baltimore: Johns Hopkins University Press.

Santer, B. D., T. M. L. Wigley, T. P. Barnett, and E. Anyamba, 1996. Detection of climate change and attribution of causes. In *Climate change 1995: Sources of climate change,* 407–43. Contribution of Working Group II to the Second Assessment of the IPCC. Cambridge, U.K.: Cambridge University Press.

Schmid, R. E., 2002. El Niño officially returns, government climate experts report. *York News-Times* (www.yorknewstimes.com/stories/071102/nat_0711020011.shtml).

Schneider, S., 1989. *Global warming: Are we entering the greenhouse century?* San Francisco: Sierra Club Books, p. 258.

Schofield, S., 1974. Seasonal factors affecting nutrition in different age groups and especially of pre-school children. *Journal of Development Studies* 11(1):22–40.

Schumacher, E. F., 1989. *Small Is beautiful.* New York: Harper Perennial.

Scott, C., 1992. The monarch butterfly. *TED Case Studies* 1(1). Washington, D.C.: American University (www.american.edu/projects/mandala/TED/butter.htm).

Sears, A. F., 1895. The coastal desert of Peru. *Bulletin of the American Geographical Society* 28:256–71.

SEHN (Science and Environmental Health Network), 1998. *Precautionary principle: Wingspread Statement.* Ames, IA: SEHN (www.sehn.org/resandac.html).

Sen, A., 1981. *Poverty and famines: An essay on entitlement and deprivation.* Oxford, U.K.: Oxford University Press.

Shapley, D., 1974. Weather warfare: Pentagon concedes 7-year Vietnam effort. *Science,* 184(4141), 1059–61.

Sharma, D., 2002. Faulty lessons from America. New Delhi: Devinder Sharma (www.dsharma.org/agriculture/faulty.htm).

Sharn, L., and M. Smaragdis, 1993. Cold puts recovery on ice. *USA Today* (16 March):A1.

Shogren, J., and M. Toman, 2000. *Climate change policy.* Discussion Paper 00-22. Washington, D.C.: Resources for the Future (www.rff.org/disc_papers/PDF_files/0022.pdf).

Simpson, S., 2001. Shrinking the Dead Zone: Political uncertainty could stall a plan to rein in deadly waters in the Gulf of Mexico. *Scientific American* 285(1):18–20 (www.sciam.com/2001/0701issue/0701scicit1.html)

Singer, F., no date. "Global warming will lower sea level rise: But will politicians listen?" *Climate Change Guest Papers* (www.vision.net.au/~daly/singer.htm).

Smith, G., [1923] 1987. Introduction. *The mystery rivers of Tibet.* Portland, Ore.: Timber Press Inc.

Smyser, W. R., 1999. *From Yalta to Berlin: The cold war strategy over Germany.* New York: St. Martin's Press.

South African Drought Commission, 1922. *Interim report of the South African Drought Investigation Commission April 1922.* Cape Town: Cape Times Ltd. Government Printers. Reprinted in *Desertification: Environmental degradation in and around Arid Lands,* ed. M. H. Glantz, 233–74. Boulder, Colo: Westview Press.

Sponberg, K., 1999. Weathering a storm of global statistics. *Nature* 400:13.

Steinbeck, J., 1945. *Cannery row.* New York: The Viking Press.

Steinbeck, J., 1939. *The grapes of wrath.* New York: Viking Penguin Books, Inc.

Stewart, T., 1997. Forecast value: Descriptive decision studies. In *Economic value of weather and climate forecasts,* ed. R. W. Katz and A. H. Murphy, 147–81. New York: Cambridge University Press.

Stolarski, R. S., and R. J. Cicerone, 1974. Stratospheric chlorine: A possible sink for ozone. *Canadian Journal of Chemistry,* 52(8):1610–15.

Suplee, C., 2000. *Milestones of science.* Washington, D.C.: National Geographic Society.

Svedin, U., 2001. *Human integration in global change research.* Summary of workshop, "Towards Integration in Global Change Research: Outlook from a Human Dimensions Perspective," held at Orsundsbro, Sweden, November 17–19, 2000. Swedish Council for Planning and Coordination of Research.

Swiss Re, 1999. El Niño 1997–98. On the phenomenon's trail. Zurich, Switzerland: Swiss Reinsurance (www.swissre.com).

Tarr, R. S., and F. M. McMurry, 1904. *A complete geography.* London: The MacMillan Co.

Taylor, G. H., and C. Southards, 1997. Long-term climate trends and salmon population Corvallis, Ore.: Oregon Climate Service (www.ocs.orst.edu/reports/climate_fish.html).

Toman, M. A., ed. 2001. *Climate change economics and policy: An RFF anthology.* Washington, D.C.: Resources for the Future.

Toynbee, A. J., 1963. *Between Oxus and Jumma.* London: Oxford University Press, p. 2.

Trager, J., 1975. *The great grain robbery.* New York: Ballentine Books.

Trenberth, K. E., J. T. Houghton, and L. G. Meira Filho, 1996. The climate system: An overview. *Climate change 1995: The science of climate change.* Contribution of Working Group 1 to the Second Assessment Report of the Intergovernmental Panel on Climate Change, ed. J. T. Houghton, L. G. Meira Filho, B. Callander, N. Harris, A. Kattenberg, and K. Maskell, 51–64. Cambridge, U.K.: Cambridge University Press.

Trenberth, K. E., G. W.Branstator, and P. A. Arkin, 1988. Origins of the 1988 North American drought. *Science* 242:1640–45.

Trewartha, G.T., 1961. *The earth's problem climates.* Madison: University of Wisconsin Press.

Tucker, C. J, H. E. Dregne, and W. W Newcomb, 1991. Expansion and contraction of the Sahara Desert from 1980 to 1990. *Science* 253:299–301.

Turco, R. P., O. B. Toon, T. P. Ackerman et al., 1984. Nuclear winter: Global consequences of multiple nuclear explosions. In *The cold and the dark: The world after nuclear war,* P. E. Ehrlich, C. Sagan, D. Kennedy, and W. O. Roberts, 163–90. New York: W. W. Norton & Co.

UNCLOS (United Nations Convention on the Law of the Sea), 1994. Agreement relating to the implementation of Part XI of the Convention. Geneva: United Nations (www.un.org/depts./los).

UNDP (United Nations Development Programme), 2001. Major donor support to UNDP for "drought proofing" initiatives in states. New Delhi: UNDP (www.undp.org.in/news/press/press207. htm).

UNEP (United Nations Environment Programme), 2002. Hotspots study sounds alarm for extinctions in the ocean: First survey to identify top ten coral reef hotspots. Nairobi: UNEP (www.unep.ch/ coral.htm).

UNFCCC (United Nations Framework Convention on Climate Change), 2001. Governments ready to ratify Kyoto Protocol. Press release (unfccc.int/cop7/press/).

UNFCCC, 2000. Climate change talks suspended: Negotiations to resume during 2001. Press release (cop6.unfccc.int/pdf/ pressreloutcome1.pdf).

UNFCCC, 1999. Ministers pledge to finalize climate agreement by November 2000. Press release (cop5.unfccc.int/media/cop5pressf.html).

UNFCCC, 1998. Kyoto Protocol talks in Buenos Aires to promote emissions cuts. Press release (cop4.unfccc.int/informed.html).

UNFCCC, 1997. Industrialized countries to cut greenhouse gas emissions by 5.2%. Press release (cop3.unfccc.int/fcc/info/indust.htm).

UNHCR (United Nations High Commissioner for Refugees), 1993. *The state of the world's refugees 1993: The challenge of protection.* New York: Penguin Books.

UNICEF (United Nations Children's Fund), 2000. Orissa emergency update: The supercyclone in Orissa in October 1999 devastated much of one of India's already poorest areas (www.unicef.org.uk/ news/emergs/orissa1.htm).

Upgren, A., and J. Stock, 2000. *Weather: How it works and why it matters.* New York: Perseus Books Group.

USA Today, 2001. New Jersey major wants investigation into snowstorm. *USA Today,* March 8 (www.usatoday.com/weather/news/ 2001/2001-0308-njsnowinvestigation.htm).

U.S. CIA (Central Intelligence Agency), 1974. *A study of climatological research as it pertains to intelligence problems.* Washington, D.C.: Library of Congress.

U.S. DOE (Department of Energy), 1985. *Characterization of information requirements for studies of CO_2 effects: Water resources, agriculture, fisheries, forests and human health.* DOE/ER-0236. Washington, D.C.: Carbon Dioxide Research Division.

U.S. House of Representatives, 1976. *A primer on climatic variation and change.* Subcommittee on the Environment and Atmosphere,

Committee on Science and Technology. Washington, D.C.: US Government Printing Office.

U.S. Senate, 1973. Senate resolution 71 (Passed 82-10). 11 July 1973. Sponsored by Sen. Pell, 93rd Congress. Washington, D.C.: Government Printing Office.

Victor, D. G., 2001. *The collapse of the Kyoto Protocol and the struggle to slow global warming.* Princeton: Princeton University Press.

Visible Earth, 2002. Satellite picture of 2002 fires and haze on Indonesian island of Borneo (www.visibleearth.nasa.gov).

Voropayev, G. V., 1997. The problem of the Caspian sea level forecast and its control for the purpose of management optimization. In *Scientific, environmental, and political issues in the circum-Caspian region,* ed. M. H. Glantz and I. S. Zonn, 105–18. NATO ASI Series No. 29. Dordrecht: Kluwer Academic Publishers.

Walmsley, D. J., and G. J. Lewis, 1984. *Human geography: Behavioral approaches,* figure 5.3, p. 52. London: Longman.

Wang, Shao-wu, 1981. *Yearly charts of dryness/wetness in China for the last 500-year period,* ed. Chinese Academy of Meteorological Science. Beijing: Map Publishing House.

Watts, M. J., 1983. *Silent violence: Food, famine and peasantry in northern Nigeria.* Berkeley: University of California Press.

WCED (World Commission on Environment and Development), 1987. *Our common future.* New York: Oxford University Press.

Webster's Ninth New Collegiate Dictionary, 1991. Springfield, Mass.: Merriam-Webster, Inc. Publishers.

Weinthal, E., 2001. *State making and cooperation: Linking domestic and international politics in Central Asia.* p. 29. Cambridge, Mass.: MIT Press.

White, G., ed. 1974. *Natural hazards: Local, national, global.* New York: Oxford University Press.

WHO (World Health Organization), 1996. *World health report 1996: Fighting disease, fostering development.* Geneva: WHO.

Whyte, A. V. T., 1985. Perception. In *Climate impact assessment (SCOPE 27),* ed. R. W. Kates, J. H. Ausubel and M. Berberia, 403–36. New York: John Wiley & Sons.

Wilson, J., 1997. Weather wars. *Popular mechanics,* February (www.popularmechanics.com/science/military/1997/2/weather_wars/print.phtml).

Windows to the Universe, 2002. Basic facts about the solar wind. Boulder, Colo.: University Corporation for Atmospheric Research,

Windows to the Universe (www.windows.ucar.edu/spaceweather/sun_earth6.html).

Witte, E., 1996. Illegal CFC trade. *TED Case Studies* 5(2). Washington, D.C.: American University (www.american.edu/projects/mandala/TED/cfctrade.htm).

Wittfogel, K., 1974. *Oriental despotism: A comparative study.* New Haven: Yale University Press.

WMO (World Meteorological Organization), 1988. *World conference on the changing atmosphere: Implications for global security.* Proceedings of workshop held in Toronto, Canada, 27–30 June 1988. WMO/OMM No. 710. Geneva: WMO.

WMO, 1984. *The global climate system: A critical review of the climate system during 1982–84.* Geneva: WMO.

Ye, Q., and M. H. Glantz, 2002. *The 1998 Yangtze floods: The use of short-term forecasts in the context of seasonal to interannual water resource management.* Report to NOAA. Boulder, Colo: National Center for Atmospheric Research.

Zonn, I. S., 2003. *Caspian encyclopedia.* Moscow: Edel-M Publishing.

ABOUT THE AUTHOR

Michael H. Glantz is a Senior Scientist in the Environmental and Societal Impacts Group, a program at the National Center for Atmospheric Research (NCAR). His research focuses on climate and society and has involved issues ranging from African drought and desertification to the development of methods of forecasting societal responses to climate change to the use of climate-related information for economic development. He received his B.S. in Metallurgical Engineering in 1961 and his Ph.D. in Political Science in 1970, both from the University of Pennsylvania. He has taught at the University of Pennsylvania, Lafayette College, University of Colorado, and Swarthmore College. He is a member of numerous national and international committees and advisory bodies related to environmental issues. In March 1990 he received the prestigious "Global 500" award from the United Nations Environment Programme. He is author and editor of several books on climate, environment, and society, including *Currents of Change: Impacts of El Nino and La Nina on Climate and Society* (second edition, Cambridge University Press, 2001) *Once Burned, Twice Shy? Lessons Learned from the 1997–98 El Niño* (United Nations University Press) and *Drought Follows the Plow: Cultivating Marginal Areas* (Cambridge, 1994), *Climate Variability, Climate Change and Fisheries* (Cambridge University Press, 1992), *Drought and Hunger in Africa: Denying Famine a Future* (Cambridge University Press, 1987 and 1988) and *Societal Responses to Regional Climatic Change* (Westview Press, 1988).

INDEX